REVIEW OF RECREATIONAL FISHERIES SURVEY METHODS

Committee on the Review of Recreational Fisheries Survey Methods

Ocean Studies Board

Division on Earth and Life Studies

NATIONAL RESEARCH COUNCIL
OF THE NATIONAL ACADEMIES

THE NATIONAL ACADEMIES PRESS
Washington, D.C.
www.nap.edu

THE NATIONAL ACADEMIES PRESS 500 Fifth Street, N.W. Washington, DC 20001

NOTICE: The project that is the subject of this report was approved by the Governing Board of the National Research Council, whose members are drawn from the councils of the National Academy of Sciences, the National Academy of Engineering, and the Institute of Medicine. The members of the committee responsible for the report were chosen for their special competences and with regard for appropriate balance.

This report is funded in part by a contract from the National Oceanic and Atmospheric Administration (NOAA). The views expressed herein are those of the author(s) and do not necessarily reflect the views of NOAA or any of its subagencies.

International Standard Book Number 0-309-10193-X

Cover: Image provided by the National Oceanic and Atmospheric Administration.

Additional copies of this report are available from the National Academies Press, 500 Fifth Street, N.W., Lockbox 285, Washington, DC 20055; (800) 624-6242 or (202) 334-3313 (in the Washington metropolitan area); Internet, http://www.nap.edu.

Printed in the United States of America

THE NATIONAL ACADEMIES

Advisers to the Nation on Science, Engineering, and Medicine

The **National Academy of Sciences** is a private, nonprofit, self-perpetuating society of distinguished scholars engaged in scientific and engineering research, dedicated to the furtherance of science and technology and to their use for the general welfare. Upon the authority of the charter granted to it by the Congress in 1863, the Academy has a mandate that requires it to advise the federal government on scientific and technical matters. Dr. Ralph J. Cicerone is president of the National Academy of Sciences.

The **National Academy of Engineering** was established in 1964, under the charter of the National Academy of Sciences, as a parallel organization of outstanding engineers. It is autonomous in its administration and in the selection of its members, sharing with the National Academy of Sciences the responsibility for advising the federal government. The National Academy of Engineering also sponsors engineering programs aimed at meeting national needs, encourages education and research, and recognizes the superior achievements of engineers. Dr. Wm. A. Wulf is president of the National Academy of Engineering.

The **Institute of Medicine** was established in 1970 by the National Academy of Sciences to secure the services of eminent members of appropriate professions in the examination of policy matters pertaining to the health of the public. The Institute acts under the responsibility given to the National Academy of Sciences by its congressional charter to be an adviser to the federal government and, upon its own initiative, to identify issues of medical care, research, and education. Dr. Harvey V. Fineberg is president of the Institute of Medicine.

The **National Research Council** was organized by the National Academy of Sciences in 1916 to associate the broad community of science and technology with the Academy's purposes of furthering knowledge and advising the federal government. Functioning in accordance with general policies determined by the Academy, the Council has become the principal operating agency of both the National Academy of Sciences and the National Academy of Engineering in providing services to the government, the public, and the scientific and engineering communities. The Council is administered jointly by both Academies and the Institute of Medicine. Dr. Ralph J. Cicerone and Dr. Wm. A. Wulf are chair and vice chair, respectively, of the National Research Council.

www.national-academies.org

COMMITTEE ON THE REVIEW OF RECREATIONAL FISHERIES SURVEY METHODS

PATRICK J. SULLIVAN (*Chair*), Cornell University, Ithaca, New York
F. JAY BREIDT, Colorado State University, Fort Collins
ROBERT B. DITTON, Texas A&M University, College Station
BARBARA A. KNUTH, Cornell University, Ithaca, New York
BRUCE M. LEAMAN, International Pacific Halibut Commission, Seattle, Washington
VICTORIA M. O'CONNELL, Alaska Department of Fish and Game, Sitka
GEORGE R. PARSONS, University of Delaware, Newark
KENNETH H. POLLOCK, North Carolina State University, Raleigh
STEPHEN J. SMITH, Bedford Institute of Oceanography, Dartmouth, Nova Scotia, Canada
S. LYNNE STOKES, Southern Methodist University, Dallas, Texas

STAFF

CHRISTINE BLACKBURN, Study Director
DAVID POLICANSKY, Scholar
SUSAN PARK, Associate Program Officer
JODI BOSTROM, Research Associate
PHILLIP LONG, Program Assistant
CARRIE WALL, Student Volunteer

Preface

The science and management of marine fisheries depend upon having clear and well-documented information. The task of collecting and maintaining this information falls to the National Marine Fisheries Service (NMFS) of the National Oceanic and Atmospheric Administration. This task is daunting given that the type and volume of information continually expand along with the needs of fisheries managers to formulate more timely and area specific management actions.

The National Research Council (NRC) has provided many fisheries and fisheries-related reviews in the last decade for Congress and NMFS. These reviews have included a summary review of the science, data, models, and processes used to guide NMFS resource management (National Research Council, 2002); an examination of how to address the legal mandate to use the best scientific information available in fisheries management (National Research Council, 2004); and a critical look at improving the collection, management, and use of marine fisheries data (National Research Council, 2000).

The current report is in response to a request from NMFS for a review of the methods used to collect and analyze recreational marine fisheries data for application to fisheries management. While recreational fisheries have long been an important component of marine fisheries resource utilization, increased fishing pressure on many stocks has heightened the demand for information from all sources. At the same time, it has become increasingly complex and challenging to assess the catch and effort associated with recreational angling.

The committee recognizes that NRC reviews add new tasks to NMFS's already hectic schedule, and we appreciate the information and responsiveness to requests that NMFS personnel provided. In particular, we thank Dr. David Van Voorhees, chief of the Fisheries Statistics Division, for his patience and openness in addressing questions about the program, and Dr. Steve Murawski, director of the Office of Science and Technology, for setting the stage for this review.

The committee also recognizes the important contribution made to this report by many individuals from regional councils, state fisheries agencies, recreational and commercial fisheries organizations, environmental conservation organizations, and others who attended and provided input to our deliberation. The people who made presentations to the committee are listed in the acknowledgments.

Finally, the committee sincerely thanks the NRC staff for their valuable support and extra efforts to facilitate the rapid completion of the report without compromising quality: David Policansky and Christine Blackburn (study directors), Susan Park (associate program officer), Jodi Bostrom (research associate), Carrie Wall (student volunteer), and Phillip Long (program assistant).

Patrick J. Sullivan, *Committee Chair*

Acknowledgments

This report was greatly enhanced by the participants of the five meetings held as part of this study. The committee would first like to acknowledge the efforts of those who gave presentations at these meetings: Allen Bingham (Alaska Department of Fish and Game), Harry Blanchette (Louisiana Department of Wildlife and Fisheries), Dick Brame (Coastal Conservation Association), Bob Bryant (recreational fisherman), Jennifer Cahalan (Washington Department of Fish and Wildlife), Felicia Coleman (Florida State University), Gordon Colvin (New York State Department of Environmental Conservation), Steve Crooke (California Department of Fish and Game), Mike Dennis (Knowledge Networks), Dave Donaldson (Gulf States Marine Fisheries Commission), Mark Fisher (Texas Parks and Wildlife Conservation Commission), Robert Fletcher (Sportfishing Association of California), Tom Fote (Jersey Coast Anglers Association), Daniel Furlong (Mid-Atlantic Marine Fisheries Council), Jeff Goebel (U.S. Department of Agriculture), John Hoey (National Oceanic and Atmospheric Administration), Cynthia Jones (Old Dominion University), Bruce Joule (Maine Department of Marine Resources), Chris Keller (Wostmann and Associates), Alec MacCall (National Oceanic and Atmospheric Administration), Herb Moore (Recreational Fishing Alliance), Steve Murawski (National Oceanic and Atmospheric Administration), Michael Nussman (American Sportfishing Association), Joe O'Hop (Florida Fish and Wildlife Conservation Commission), Vince O'Shea (Atlantic States Marine Fisheries Commission), Maury Osborn (Atlantic Coastal Cooperative Statistics Program), Preston Pate (North Carolina Division of Marine Fisheries), Bonnie Ponwith (National Oceanic and Atmospheric Administration), Russell Porter (Pacific States Marine Fisheries Commission), Robin Reichers (Gulf of Mexico Fisheries Management Council), Ronald Salz (National Oceanic and Atmospheric Administration), Eric Schindler (Oregon Department of Fish and Wildlife), Richard Stone (private fisheries consultant), Cynthia Thomson (National Oceanic and Atmospheric Administration), David Van Voorhees (National Oceanic and Atmospheric Administration), Bobbi Walker (Gulf of Mexico Fisheries Management Council and National Association of Charterboat Operators), and Robert Zales II (National Association of Charterboat Operators). These talks helped set the stage for fruitful discussions in the closed sessions that followed. The committee is also grateful to the many people who provided important discussion during the public comment periods.

This report has been reviewed in draft form by individuals chosen for their diverse perspectives and technical expertise, in accordance with procedures approved by the NRC's Report Review Committee. The purpose of this independent review is to provide candid and critical comments that will assist the institution in making its published report as sound as possible and to ensure that the report meets institutional standards for objectivity, evidence, and responsiveness to the study charge. The review comments and draft manuscript remain confidential to protect the integrity of the deliberative process. We wish to thank the following individuals for their participation in the review of this report:

John Annala, Gulf of Maine Research Institute, Portland, Maine
James R. Chromy, RTI International, Research Triangle Park, North Carolina
John Mark Dean, University of South Carolina, Columbia
John E. Halver, University of Washington, Seattle
David K. Loomis, University of Massachusetts at Amherst
Kenneth E. McConnell, University of Maryland, College Park
Jean Opsomer, Iowa State University, Ames
Maury Osborn, Atlantic Coastal Cooperative Statistics Program, Washington, DC
Russell Porter, Pacific States Marine Fisheries Commission, Portland, Oregon
Norman H. Sleep, Stanford University, California
Bobbi Walker, National Association of Charterboat Operators, Orange Beach, Alabama

Although the reviewers listed above have provided many constructive comments and suggestions, they were not asked to endorse the conclusions or recommendations nor did they see the final draft of the report before its release. The review of this report was overseen by **Andrew Solow**, Woods Hole Oceanographic Institution, appointed by the Divison on Earth and

Life Studies, and **John Dowling**, Harvard University, appointed by the Report Review Committee, who were responsible for making certain that an independent examination of this report was carried out in accordance with institutional procedures and that all review comments were carefully considered. Responsibility for the final content of this report rests entirely with the authoring committee and the institution.

Contents

Summary

INTRODUCTION

Recreational fishing in the United States is an important social and economic component of many marine fisheries. However, in some cases, recreational fishing takes more fish than commercial fishing, and in an increasing number of cases, recreational fishing is the main source of fishing mortality. In addition, current assessments indicate that some marine recreational fisheries have exceeded their quotas, raising concern because fishing effort in marine recreational fisheries is projected to increase. It is important that catch monitoring systems are adequate for timely management of these fisheries.

Marine recreational fisheries are not monitored with the same rigor as commercial fisheries. However, as concerns about the effects of all types of fishing have grown, more attention has been paid to the possible impacts of marine recreational fishing. The growing interest in the effects of recreational fishing on fish stock size and composition has led to increased demands for timely and accurate data. Although the National Marine Fisheries Service (NMFS) of the National Oceanic and Atmospheric Administration implemented the Marine Recreational Fisheries Statistics Survey (MRFSS) in 1979 to obtain statistics about marine recreational fisheries, management goals and objectives have changed since then, as has the complexity of the recreational fishing sector. The need for and use of marine recreational fishery statistics in science and management have changed as well. This committee has identified several areas in which designers of sampling programs, data collectors, and users of recreational fisheries data appear to have incomplete communication, mismatched criteria, or other obstacles.

The MRFSS has two major components: an onsite component, in which anglers are intercepted and interviewed on the water or at sites such as marinas where they access the water; and an offsite component, in which anglers are contacted and surveyed by telephone after their trips are completed. There has been widespread criticism of the nature and use of the MRFSS information. The MRFSS was (and is) intended to be a national program, but not all coastal states participate. In some cases, states have their own surveys of recreational fish landings instead of the MRFSS; in other cases, states have surveys that complement the MRFSS. In addition to this lack of uniformity of coverage, the quality of the MRFSS data for management purposes has also been questioned.

Indeed, it is much more difficult to collect data on recreational saltwater anglers than on commercial fishing operations. There are far more saltwater anglers than commercial fishermen—approximately 14 million anglers fished annually in recent years—and they do not land their catches at specific points where there are dealers, as do commercial fishermen. In addition, there are many modes of fishing (e.g., anglers who fish from head boats or charter boats, with guides,[1] from shore, on private boats, from private property), and many anglers release fish they catch. Some anglers travel far to fish and often fish only a few times each year, which makes them difficult to encounter in surveys. Others, who live within 50 miles of the coast, are much more likely to be intercepted by the MRFSS. Finally, most surveys of anglers depend to some degree on the anglers' recall and willingness to volunteer valid information. As a result, designing a survey that will provide accurate and timely information, with good coverage and at acceptable cost, is a major challenge.

Despite the complexity of the challenge and its importance for fishery management, the MRFSS staff have been severely handicapped in their efforts to implement, operate, and improve the MRFSS, including implementing the recommendations of earlier reviews. It is not reasonable to expect such a small staff—and one that lacks a Ph.D.-level mathematical statistician—to operate a national survey of such complexity, despite the dedication of the small staff the MRFSS does have.

[1] Head boats, also called party boats, take large groups of anglers (sometimes as many as 100) on fishing trips; the groups usually are not pre-formed. Charter boats (also occasionally called party boats) take smaller groups of anglers, usually four to eight, most often in pre-formed groups. Guided trips are trips in which a guide takes one or two anglers in a smaller boat. These different categories operate under different U.S. Coast Guard and state license requirements. Throughout this report, these sectors are collectively referred to as the for-hire sector.

In addition, the MRFSS is severely limited by the lack of a universal sampling frame for all saltwater anglers, a lack that is not of the MRFSS's own making. To make matters even more difficult, some of the data that the MRFSS depends on are collected by states, which use a variety of data-collection and sampling protocols. Finally, the financial resources allocated to the MRFSS are modest in comparison to the challenge. This committee's findings and recommendations should be viewed with this in mind.

THE PRESENT STUDY

To help identify solutions to some of the above problems, NMFS asked the National Academies to assemble a committee to review current marine recreational fishing surveys and to make recommendations for improvements—especially to the MRFSS—and to recommend the implementation of possible alternative approaches. (See Box S.1 for the committee's statement of task.)

In response, the National Research Council (NRC) of the National Academies established the Committee on the Review of Recreational Fishing Survey Methods, composed of experts in survey design and statistics, biological statistics, fishery management, and the economics and sociology of recreational fishing. The background and support for the conclusions and recommendations presented below are found in subsequent chapters.

CONCLUSIONS AND RECOMMENDATIONS

General Conclusions

- The committee agrees with conclusions of previous NRC committees that marine recreational fishing is a significant source of fishing mortality for many marine species and that adequate scientific information on the nature of that mortality in time and space is required for successful management of those species.
- Marine fisheries management goals, objectives, and context have changed since the MRFSS was begun in 1979. Management

Box S.1
Statement of Task

This study will critically review the types of survey methods used to estimate catch per unit effort and effort in recreational fisheries, including state and federal cooperative programs. The committee will examine representative survey types but will not evaluate every regional or state survey method currently in use. The study will consider the match or mismatch between options for collecting recreational fisheries data and alternative approaches for managing recreational fisheries.

In particular, the committee will assess current types of survey methods giving consideration to:

- The suitability for monitoring different types of fishing (e.g., charter boats versus private boats, offshore versus near-shore species, fisheries with temporally or spatially restricted fishing seasons).
- The adequacy for providing the quality of information needed to support various approaches for managing recreational fisheries, with reference to how the management approach might be restricted by the type of survey method, stratification scheme, and sample size required. For example, is the management time frame (in-season, annual, or multi-year) consistent with temporal design of the survey? Is the geographic scale of management (e.g., state versus regional) appropriate for the resolution provided by the survey? How would the survey design need to be modified to match the requirements of the management approach?
- Make recommendations regarding possible improvements to current surveys and/or possible implementation of alternative approaches, including setting priorities for revising monitoring methods that will yield the greatest improvements in effort and catch per unit effort estimates.

Current survey methods and recommended alternatives will be compared with relation to costs, sources of bias, precision, and timeliness.

decisions are often made at finer spatial and temporal scales than they were earlier, the mix of recreational and commercial fishing has changed for many areas and species, and stock-assessment models now make greater use of data from recreational fisheries.

- The MRFSS is in need of additional financial resources so that technical and practical expertise can be added to assist in a major overhaul of the design, implementation, and analysis of data from the MRFSS. Both the telephone and access components of the current approach have serious flaws in design or implementation and use inadequate analysis methods that need to be addressed immediately.

- This committee's review has focused primarily on the MRFSS, but many of the component surveys of the MRFSS conducted by state agencies (with various degrees of federal funding) suffer from the same shortcomings as does the central MRFSS. As a result, many of this committee's recommendations apply to state surveys as well as to the MRFSS.
- Many of the independent surveys conducted by the states, as well as state-run surveys that are components of the MRFSS, are different from each other and from the central MRFSS in important ways, including sampling, data collection, and preparation of estimators.
- The committee concludes that users' concerns about the use of the MRFSS in fishery management are justified by the above-mentioned weaknesses, but they also result from inadequate communication and outreach on the part of the MRFSS managers at NMFS.
- The for-hire sector of marine recreational fisheries (i.e., charter, guide, and head boat operations) is more like a commercial sector than it is like the private-angler sector.

General Recommendations

- The MRFSS (as well as many of its component or companion surveys conducted either indirectly or independently) should be completely redesigned to improve its effectiveness and appropriateness of sampling and estimation procedures, its applicability to various kinds of management decisions, and its usefulness for social and economic analyses. After the revision is complete, provision should be made for ongoing technical evaluation and modification, as needed, to meet emerging management needs. To improve the MRFSS, the committee further recommends that the existing MRFSS program be given a firm deadline linked to sufficient program funding for implementation of this report's recommendations.
- A much greater degree of standardization among state surveys, and between state surveys and the central MRFSS, should be

achieved. This will require a much greater degree of cooperation and coordination among the managers of the various surveys.

- The for-hire sector of marine recreational fisheries should be considered a commercial sector, and survey methods and reporting requirements for that sector therefore should be different from those for private anglers.

Sampling Issues Conclusions

- The committee concludes that the current methods used in the MRFSS for sampling the universe of anglers and for determining their catch and effort are inadequate. Sampling of each group of anglers (i.e., private, guided, head boat, charter boat) presents challenges that can differ across the groups. Two complementary methods of sampling are used in the MRFSS. One is onsite (i.e., intercepting anglers while they are fishing or at their access [landing] points). The other is offsite, which includes a variety of sampling techniques for contacting anglers after they have completed their trips. Both onsite and offsite methods suffer from weaknesses that may lead to biases in catch and effort estimation. Finally, the estimation procedure for information gathered onsite does not use the nominal or actual selection probabilities of the sample design and therefore has the potential to produce biased estimates for both the parameters of interest and their variances.
- Onsite methods fail to intercept anglers who have private access to fishing waters or intercept them only sporadically. It is impossible, using current methods, to obtain information on the target species of anglers who have private access. In addition, various physical, financial, and operational constraints often lead to spatial or temporal biases in onsite sampling coverage that are not adequately accounted for in the estimation equations.
- Offsite sampling methods that rely on telephone interviews are complicated by the increasing use of cellular telephones, especially in surveys of residents of coastal counties. This is because cellular telephones are not restricted to a geographic region as are landline telephones. If cellular telephones are excluded, then undercoverage of the survey will be increasingly problematic

over time as the number of people who use only cellular telephones is growing.

- The existing random digit dialing (RDD) survey suffers in efficiency from the low proportion of fishing households among the general population and may allow bias in estimation from its restriction to coastal counties only.
- Reliance on fishing license-based lists of saltwater anglers is not yet feasible as a means of improving offsite sampling methods to avoid the inefficiency of RDD, undercoverage due to cellular telephone use, and restriction to coastal counties. Although many states collect angler information when a saltwater fishing license is purchased, there are license exemptions based on age, residence, access points, existence of a boat license, mode of fishing, and other factors. As a result, angler information for those states is incomplete. Some states have more complete information than others, and in the states that have no saltwater license, there is no list of saltwater anglers. The lack of a universal sampling frame (registry or license requirement) for all saltwater anglers is a major impediment to the development of a reliable and accurate survey program.
- Catch and release fishing (release of fish that survive capture) is increasingly common in many marine recreational fisheries. Although some fish survive capture and release, mortality may be high, in some cases exceeding 50 percent. The survey fails to provide a valid and reliable method of adequately accounting for fish caught and *not* brought to the dock (including fish released alive or dead, as well as fish caught for bait or given away before reaching the dock). This shortcoming affects estimates of catch and total removals.
- The correct identification of fish species, especially in places with diverse fish faunas, is a difficult challenge, both for many anglers and for those conducting surveys. Incorrect identification obviously has the potential to lead to incorrect conclusions from survey data.

Sampling Issues Recommendations

- A comprehensive, universal sampling frame with national coverage should be established. The most effective way to achieve this is through a national registration of all saltwater anglers or through new or existing state saltwater license programs that would allow no exemptions[2] and that would provide appropriate contact information from anglers fishing in all marine waters, both state and federal. Any gaps in such a program (e.g., a lack of registration in a particular region or mode, exemptions of various classes of anglers) would compromise the use of the sampling frame and, hence, the quality of the survey program. An updated, complete registration list would greatly improve sampling efficiency in terms of time and cost. Although these savings might not cover the entire cost of maintaining such a database, the benefit from the increased quantity and quality of the data would be worth the extra cost, especially if there is an associated increase in public confidence in the final estimates.

- Future telephone surveys should be based on the above universal sampling frame.
- Charter boat, head boat, and other for-hire recreational fishing operations should be required to maintain logbooks of fish landed and kept, as well as fish caught and released. Providing the information should be mandatory for continued operation in this sector, and all the information should be verifiable and made available to the survey program in a timely manner.
- Additional studies are needed to understand the extent to which fish are kept and inspected, as well as the extent of catch not available for inspection to improve the accuracy of catch estimates.
- Panel surveys, which contact individual anglers repeatedly over time, should be considered in recreational fishing surveys to gather angler trend data and to improve the efficiency of data collection.
- The onsite sampling frame for the MRFSS should be redesigned. The estimation procedure critically depends on the assumption that catch rate does not vary according to the nature of the access point. In particular, small or private access points that most likely are missed might have different catch rates than larger access points, which would lead to bias in the resulting estimators. In

[2] There is no scientific reason that a state should not continue to allow certain groups (e.g., seniors) to fish for free, as long as everyone is required to register in the universal sampling frame or have a state saltwater license.

addition, the sampling process requires greater quality control (less latitude on the part of the samplers) than it has at present. (See the recommendation below for the establishment of an independent research group to investigate matters such as these.)

- Dual-frame procedures should be used wherever possible to reduce sample bias. For example, if a state has an incomplete list frame based on licenses, the use of an additional sampling frame of the state's residents (e.g., RDD) would reduce the bias. The existence of a universal frame described above would make this approach unnecessary for offsite sampling.
- Internet surveys should be considered for their potential use in recreational fishing surveys, especially in panel surveys, as a way for anglers to submit information.

Statistical Estimation Issues Conclusions

- The designs, sampling strategies, and collection methods of recreational fishing surveys do not provide adequate data for management and policy decisions. Unknown biases in the estimators from these surveys arise from reliance on unverified assumptions. Unless these assumptions are tested and the degree and direction of bias reliably estimated, the extent to which the biases affect final estimates will remain unknown.
- The statistical properties associated with data collected through different survey techniques differ and often are unknown. The current estimators of error associated with various survey products are likely to be biased and too low. It is necessary, at a minimum, to determine how those differences affect survey results that use differing methods.

Current analysis procedures used in the MRFSS do not exploit the current knowledge of finite population sampling theory. The current estimates are particularly deficient when applied to small areas because they do not use information in adjoining areas or time periods, nor do they consider relationships between species that occur together. Therefore, they are of lower precision than would be possible if this information were used. Improvements in these estimates would be of great use to managers who need to make quick decisions concerning spatial areas that are smaller than typical in the early years of the MRFSS.

Statistical Estimation Issues Recommendations

- The statistical properties of various sampling, data-collection, and data-analysis methods should be determined. Assumptions should be examined and verified so that biases can be properly evaluated.
- A research group of statisticians should design new analyses based on current developments in sampling theory. These examinations should include experimentation, such as specific sampling of activities like nighttime fishing or fishing from private property, whose current underrepresentation in the MRFSS sampling has the potential to create bias.

Human Dimensions Conclusions

- The MRFSS was not designed with human dimensions data (i.e., collection of social, behavioral, attitudinal, and economic data) in mind. The qualities of social, economic, and other human dimensions data have been compromised for many of the same reasons that the biological data have been compromised, including such issues as those related to coastal populations, telephone surveys, and sampling protocol. The human dimensions data have been further compromised by simply being added onto the biological data collection efforts that have different sampling requirements and survey design needs. Current surveys are largely focused on biological factors (e.g., numbers, sizes, and species of fish landed) and not on human dimensions factors. The statistical and sampling problems associated with social, behavioral, attitudinal, and economic data often can be considerably different from those associated with biological factors.

  If the number of marine fishing trips increases, it is likely that additional fishing access sites will be developed. In addition, social and environmental changes (e.g., changes in the distribution and numbers of people, a major hurricane) also can affect the availability and use of access sites. To ensure adequate coverage of the recreational fishery, a periodic updating of lists and descriptions of fishing locations and access sites is needed.

Human Dimensions Recommendations

- An independent national trip and expenditure survey should be developed to support economic valuation studies, impact analyses, and other social and attitudinal studies. The sampling and survey procedures of the independent survey should be designed for the purpose of social and economic, not biological, analyses.
- Add-on surveys for human dimensions should be continued but in a more focused way than currently is done to target specific management needs and to supplement the national data as needed.
- The national database on marine recreational fishing sites and their characteristics should be enhanced to support social, economic, and other human dimensions analyses. Sites should be defined at levels as fine as possible. The data set should include site characteristics that matter to anglers in making fishing choices, such as boat ramps, facilities, natural amenities, parking, size, and type (e.g., beach, pier, launch point). To account for changes in the number and patterns of trips and the changing characteristics of sites, a periodic updating of the data should be conducted.

Program Management and Support Conclusions

- A large number of complex technical issues associated with surveys of marine recreational fishing remain unsolved, and a significant investment in intellectual and technical expertise is needed.
- A greater degree of coordination between federal, state, and other survey programs is necessary to achieve the national perspective on marine recreational fisheries that is needed.
- The recommended changes to the design and operation of the MRFSS and its continued development and operation will require additional funding above current levels.

Program Management and Support Recommendations

- A permanent and independent research group should be established and funded to continuously evaluate the statistical design and adequacy of recreational fishery surveys and to guide necessary modifications or new initiatives. Human dimensions expertise should be included as well.
- Additional funding is needed for a survey office devoted to the management and implementation of marine recreational surveys, including coordination between surveys conducted in various state and federal agencies.

Communication and Outreach Conclusions

- It is difficult for individual anglers to see the effects of recreational fishing on their target species and to distinguish daily and seasonal fluctuations from trends. As a result, no matter how well designed and implemented a marine recreational survey is, it will not fully succeed without the cooperation of anglers. Unless anglers believe that the survey is well designed and implemented and that it is being used intelligently to address appropriate management issucs, they are unlikely to participate.
- In particular, anglers need to have a basic understanding of the relationship between a statistically based sampling scheme and the frequency with which each of them is (or is not) contacted by a data collector.
- If anglers believe that their input is influencing the design and use of surveys, they are more likely to be satisfied with those surveys than otherwise.
- If anglers understand the basic purposes and decisions to which recreational fishing survey data are being applied and how those data are interpreted and used, they are more likely to feel confident that the approaches used are legitimate and are more likely to participate willingly and provide valid information.

Communication and Outreach Recommendations

- Outreach and communication should be improved in several ways. The MRFSS managers should advise anglers and data users on the constraints that apply to the use of the data for various purposes. Managers and anglers also should be informed clearly about any limitations of the data.

- Outreach and communication should be institutionalized as part of an ongoing MRFSS program so their importance is acknowledged and appropriate expertise can be developed.
- Angler associations should be engaged as partners with survey managers through workshops, data collection, survey design, and participation in survey advisory groups. Many NRC and other reports stress the importance of using local and traditional knowledge, capacity building, and local communities in knowledge-gathering and dissemination activities. These recommendations apply, as well, to the recreational fishing community.

1

Introduction

Recreational fisheries are an important and growing component of many marine fisheries. Data about the numbers and kinds of fish taken through recreational fishing are essential for fisheries management. In recent years, recreational fisheries have been monitored less rigorously than commercial fisheries. But as concerns about overfishing have grown, more attention has been turned toward the possible impact of marine recreational fishing and the proportion of fish taken by each sector (e.g., National Research Council, 1999, 2000; Lucy and Studholme, 2002; Coleman et al., 2004). This, in turn, has led to greater demands for timely, accurate recreational fishing data and scrutiny of the methods used to collect these data.

According to the best available estimates, approximately 14 million anglers made almost 82 million fishing trips along the Atlantic, Pacific, and Gulf coasts (excluding Texas) in 2004 (National Oceanic and Atmospheric Administration, 2005a). While each individual angler typically harvests a small number of fish, collectively these sport fisheries can take a significant fraction of the yearly catch. For example, in 1999, recreational fishing accounted for 94 percent of the total catch of spotted seatrout (*Cynoscion nebulosus*), 76 percent of striped bass (*Morone saxatilis*) and California sheephead (*Semicossyphus pulcher*), and 60 percent of king mackerel (*Scomberomorus cavalla*) (Figure 1.1). Recreational catch continues to grow for several important fisheries; the recreational catch of summer flounder and Pacific halibut increased 40 fold in less than 20 years (Coughenower and Blood, 1997). Recent scientific papers (e.g., Coleman et al., 2004; Cooke and Cowx, 2004) provide data that suggest recreational fisheries are a more significant factor in the exploitation of fish stocks than previously believed. Considering the potential contribution recreational fishing has to total

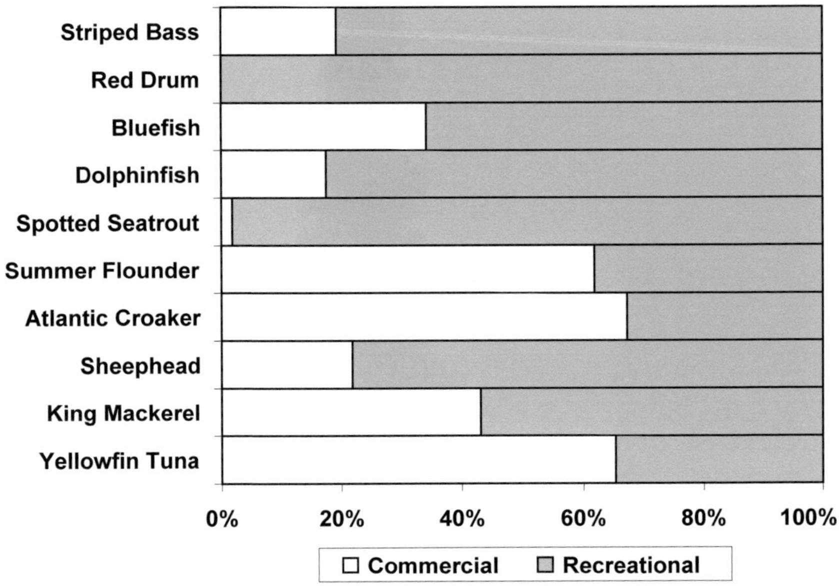

FIGURE 1.1 Top Ten Recreational Species Versus Commercial Harvest for 2004. Comparisons between the top ten species in descending order of abundance by weight for U.S. recreational fish harvests and commercial landings. The figure does not include data for Alaska and Texas because no NMFS recreational surveys are conducted in those states (National Oceanic and Atmospheric Administration, 2005a).

catch, it is clear that accurate monitoring of catch from both recreational and commercial fisheries is needed to ensure that total catch does not exceed the total allowable catch calculated to maintain a sustainable population.

Nationally, recreational catch is monitored primarily (but not entirely) through the Marine Recreational Fisheries Statistics Survey (MRFSS), which was implemented by the National Marine Fisheries Service (NMFS) of the National Oceanic and Atmospheric Administration (NOAA) in 1979 to obtain standardized and comparable estimates of participation, effort, and catch by anglers in U.S. marine waters (Essig and Holliday, 1991; National Oceanic and Atmospheric Administration, 2005b). In addition, there are several state programs, operated instead of or complementary to the MRFSS, which are discussed in Chapter 3. The stated purpose of the MRFSS is to establish a reliable database for estimating the impact of marine recreational fishing on marine resources. According to the MRFSS, "[t]he program's mission is to provide accurate, precise, and timely fisheries-dependent information for U.S. marine waters through the coordination and administration of recreational fisheries surveys nationwide" (National Oceanic and Atmospheric Administration, 2005b). However, many now argue that recreational data collected through the MRFSS and other recreational fishing surveys are being used for management decisions in ways that exceed their intended design and purpose (Box 1.1).

As exploitation levels of the nation's fisheries have increased, managers have begun (of necessity) to manage stocks more actively. This has resulted in the close monitoring of catch; more frequent closings of fishery areas and shortening of seasons; more stringent size limits; and for anglers, more detailed, species-specific bag limits. Compared to only 10 years ago, recreational fisheries data are now incorporated into more fisheries management plans and stock assessments of the regional fishery management councils. Currently, many recreational fisheries are managed on annually monitored quotas, and some are even managed using in-season quotas. This type of recreational fisheries management requires data to be collected and analyzed in the span of weeks instead of months—the amount of time it has traditionally taken to collect and process recreational data. Other recreational fisheries require monitoring and management on finer geographical scales. For example, the quota for the striped bass stock on the east coast is divided among states, and each state is required to report its respective catch. This is also true for king mackerel.

As these new needs have arisen, some of the challenges have been met with changes to the MRFSS, and in some cases, new surveys have been implemented to provide some of the necessary data (see Appendix C). However, concern remains that the data currently being collected through recreational fishing surveys are not precise, robust, or timely enough for the various scales employed in fisheries management. The mismatch between the data required and the data delivered has prompted substantial commentary on and criticism of the program. In addition, there are occasions when the estimates of total catch vary widely between years, leading to significant changes in the management restrictions implemented, again eliciting questions and criticisms from recreational anglers and commercial fishermen.

The data produced by these surveys are subjected to considerable scrutiny, but there is a lack of confidence within some elements of the angler community regarding the accuracy of total removal estimates of

Box 1.1
Users of the MRFSS Information

The MRFSS information is used by a variety of different groups. The Magnuson Fishery Conservation and Management Act of 1976, amended in 1980, and now the Magnuson-Stevens Fishery Conservation and Management Act (P.L. 94-265) gives much fishery management responsibility to eight regional fishery management councils. In addition, there are three marine fisheries commissions—the Atlantic States Marine Fisheries Commission, the Gulf States Marine Fisheries Commission, and the Pacific States Marine Fisheries Commission—which have various degrees of management and regulatory authority. There are four international commissions that make management recommendations to their governments, including the U.S. government, for implementation: the Pacific Salmon Commission (United States and Canada), the International Pacific Halibut Commission (United States and Canada), the Inter-American Tropical Tuna Commission (United States and 14 other member countries), and the International Commission on Conservation of Atlantic Tunas (United States and 40 other member countries). (The Great Lakes Fishery Commission [United States and Canada] does not take part in or use the MRFSS.) The commissions manage species that are recreationally important. The coastal states have fishery management agencies with jurisdiction in state waters (usually to three nautical miles offshore but nine nautical miles for Texas, Florida, and Puerto Rico). Other users include fishing organizations, environmental groups, various industry groups, and individuals. All these groups use data from the MRFSS and other sources to make decisions concerning their various activities.

recreational fish catch. This has resulted in reduced angler cooperation with the data-collection surveys (because participation is voluntary) and in disagreements about proposed management actions. On some occasions, these disagreements have resulted in lawsuits against NMFS. Some user groups have even generated alternate surveys in attempts to illustrate shortcomings of existing surveys or to provide a more appropriate design for local circumstances.

This report will demonstrate that achieving valid and reliable recreational fishing estimates in the future will require much better cooperation between agencies; the redesign of some existing surveys; the creation of new surveys; and, most important, increased funding to make the above practicable.

WHAT DATA ARE COLLECTED FROM ANGLERS

Usually, recreational surveys at large spatial scales measure angler effort (E) using one survey (e.g., telephone survey) and measure catch per unit effort (CPUE) using a separate survey (e.g., angler interview). If E and CPUE are known, one can calculate total catch (C) as:

$$C = E \times CPUE$$

There are three different categories of catch: catch available for inspection during the onsite interview (A); catch unavailable for inspection because it is filleted, discarded dead, or refused for inspection (B1); and catch released alive (B2). For each angler trip, total catch is defined as A + B1 + B2, and landed catch, or harvest (H), is calculated as A + B1. For management purposes such as quota monitoring, the total number of fish taken by anglers is recorded as harvest; catch released alive is not factored in.

For stock assessment purposes, an additional fraction of the catch released alive that subsequently dies is estimated using a "hooking" mortality (M_H). This enables the estimation of total recreational fishing removals (R):

$$R = A + B1 + (B2 \times M_H)$$

Catch, harvest, and total removals are estimated by species for specific spatial regions and temporal periods, with the spatial and temporal scale requirements depending on the management needs for that species. Further, catch is assigned to age or size classes based on biological sampling of the catch.

HOW CATCH IS ESTIMATED FROM RECREATIONAL FISHERIES SURVEY DATA

The data produced by surveys of anglers are used to provide information for stock assessments and to support management decisions. An accurate estimate of total removals is the most critical piece of information needed for marine fisheries management (Gulland, 1983; Pauly, 1998; Watson et al., 2000), but other estimates also are required, including effort expended; species targeted; geographical allocation of landings; and the species, age, and size composition of the catch. In addition to quota monitoring and stock assessments, recreational data are also important for social and economic studies that help to determine current socioeconomic impacts of fishing participation and help to evaluate the effectiveness of current and potential fishery management actions. However, economic and social data generally are not collected in the same consistent, annual approach as catch and CPUE data.

WHY RECREATIONAL FISHING DATA ARE DIFFICULT TO COLLECT

While there are problems with commercial catch statistics, such as lack of mandatory reporting requirements for some fisheries and incorrect reporting, in general, it is more straightforward and less complex to collect catch statistics for the commercial sector than the recreational one. Commercial fisheries catches are usually landed at specific points where dealers are located. At these places, catch is weighed and recorded, providing an opportunity to monitor this catch through dockside landing receipts. Since both the dealer and the fisherman record and submit the data collected, there is a useful redundancy in the data that facilitates error checking and quality control. More specific effort and discard information is often documented through mandatory logbooks and observer programs. Furthermore, the universe of known fishermen in the commercial sector is defined through required commercial licenses. All of these facts align to provide relatively reliable and timely estimates from the commercial sector.

Conversely, the participation base for the recreational fishery is considerably more numerous, diverse, and diffuse than that for commercial fisheries. In addition, anglers access fisheries in numerous modes[1] so there are not limited access points analogous to processing facilities at which to intercept anglers. The nature of the participants is also different. Many anglers are infrequent participants in the fishery and may be less amenable to survey coverage.

The goal of monitoring all different modes, locations, and species involved (e.g., charter boats versus private boats, offshore versus near-shore species, fisheries with temporally or spatially restricted fishing

---

[1] The type of place or platform from which marine recreational fishing occurs, including fishing from boats sailing from harbors, marinas, and private docks and from shore on piers, jetties, and beaches.

seasons) is a lofty one. Indeed, recreational surveys may be the most complex national surveys currently conducted. With so many complexities, accurate estimation of total catch is very expensive. Current methods circumvent some of this expense by making unverified—and, in some cases, unverifiable—assumptions about angler behavior. The extent to which these assumptions adversely affect the estimates is not currently known and will need to be evaluated.

The When and Where of Data Collection

The current management framework for marine recreational fisheries is highly diverse, depending on species and region. The temporal spectrum of management actions runs from in-season to annual to multi-year, all of which require different data collection and processing strategies. In-season management obviously requires the most timely data provision to avoid quota overruns. In fact, this approach has been implemented previously for some species and then abandoned due to the lack of estimates of sufficient precision at the appropriate temporal and spatial scale (e.g., red snapper in the Gulf of Mexico).

The spatial scale for management extends from local to regional[2] and national to, occasionally, international. In addition, species may be managed individually or jointly with co-occurring species. The adequacy of a particular survey design for supporting a particular temporal scale of management (in-season, annual, or multi-year) may require a similarly resolved spatial scale. As finer management scales are created, sampling strata that were previously combined to create a mean estimate on a larger scale must now stand alone. This means a larger sample size from each stratum is needed to achieve reasonable precision of the estimates within each stratum. This requires evaluation and revision on a continuing basis.

---

[2] The committee uses the term "regional" to represent scales that are smaller than national but larger than state, and consequently, there will be some association of this scale with the jurisdictions of regional councils and multistate commissions. In some instances, however, the context could be interpreted more broadly.

The Who of Data Collection

Recreational fisheries surveys involve the interaction of two large and diverse groups of people: the fishing community and the survey organizations. The recreational fishing community can be usefully divided into two populations based on the mode of fishing. The first includes anglers who fish from shore, docks, and private (i.e., not for-hire) boats. The second, the for-hire sector, is comprised of anglers who fish from head boats, charter boats, and guided boats.[3] The issues that impact sample design are different for the two populations. They differ in total numbers of participants, the ability to access the anglers for interview, the frequency of fishing activity, and the skill of each type of angler. Typically, these two populations are sampled separately with the for-hire sector often sampled using a boat directory telephone survey.

Organizations tasked with gathering the data for these angler surveys vary across regions and states; this can contribute to variations in the quality and quantity of the data collected. NMFS conducts the MRFSS through contracts to private companies that complete the telephone and intercept interviews. Some state agencies have taken on the responsibility of conducting intercept interviews, using state personnel to collect the catch data for each angler, while contractors continue to conduct the telephone interviews. Still other states carry out all aspects of survey implementation (e.g., Texas, California, Oregon, Washington); in California, Oregon, and Washington, the survey activities receive partial funding from NMFS. The degree to which the regional fisheries commissions are involved in survey implementation and data collection varies as well. The Gulf States Marine Fisheries Commission serves as the contractor for the intercept portion of the MRFSS in that area; whereas the Atlantic States Marine Fisheries Commission is involved in survey design and data handling for its area but is not involved in the actual conduct of the data collection. The Pacific States Marine Fisheries Commission (PSMFC) coordinates sampling in California, Oregon, and

[3] Head boats, also called party boats, take large groups of anglers (sometimes as many as 100) on fishing trips; the groups usually are not pre-formed. Charter boats (also occasionally called party boats) take smaller groups of anglers, usually four to eight, most often in pre-formed groups. Guided trips are trips in which a guide takes one or two anglers in a smaller boat. These different categories operate under different U.S. Coast Guard and state license requirements. Throughout this report, these sectors are collectively referred to as the for-hire sector.

Washington between the states and NMFS, with a NMFS contract and state funds. The sampling programs are a mix of state and PSMFC samplers and supervisors. PSMFC maintains the catch and effort database for the three states with the online Recreational Fisheries Information Network (RecFIN). RecFIN is used by the states and the Pacific Fishery Management Council to monitor and manage the fisheries under their purview. The RecFIN database is also available for access by recreational fisheries constituents and the public. Clearly, agency cooperation and coordination is crucial to producing consistent and useful data and estimators.

The Cost of Data Collection

The amount of data collected is limited most often by cost. Each telephone call made and each intercept survey conducted has a specific price tag associated with it depending on labor costs of the surveyor and the amount of training that person needs to conduct the survey properly. In 2004, the annual operating budget for the MRFSS was approximately $7.2 million. In addition, many states have chosen to supplement existing surveys with additional funds, increasing the overall sample size to achieve specific state goals. For example, North Carolina dedicates approximately $260,000 per year in additional funds to supplement the MRFSS sampling. Most often, additional money is expended because of the need for more accurate or precise estimates to better manage the resources. The cost of the data collection should always be weighed against the benefit of having more information. This committee cannot judge how much federal money should be spent on any endeavor, but one important aspect of the research it recommends in Chapter 6 would be to identify the financial costs and benefits in terms of various improvements in the survey.

SCIENTIFIC ISSUES AFFECTING SURVEY DESIGN

For all surveys, there are concerns that must be addressed adequately to foster confidence in the generated estimates. Primary among these are the statistical concerns of precision and bias. (See Box 1.2 for a few relevant statistical definitions; see Chapter 3 for more detailed discussion of these statistical issues.) Inadequate precision can be addressed in a straightforward manner by increasing sample size or by otherwise increasing the efficiency of the sample design. Bias is more difficult to identify and reduce. It typically cannot be measured from the sample itself and cannot be reduced without changes in the way the sample units are selected or the way measurements are taken from those units.

In an ideal survey, the target population units are identified in a frame, a sample is selected from the frame, and the selected units respond with accurate information. Unfortunately, errors can arise at each step in this process. Groves et al. (2004) classify errors as either errors of representation or errors of measurement.

Errors of representation are those that arise due to problems that prevent the sample from representing the population accurately. These errors can lead to bias in the estimates if the excluded population units differ from the included ones. In fisheries surveys, these errors include the following:

- Coverage error occurs when the sampling frame does not match the target population perfectly, due to duplications or under-coverage. Duplications occur when the frame lists a target unit more than once, such as a fishing household with more than one telephone line. Undercoverage occurs when the sampling frame does not include all the units belonging to the target population. This occurs when not all sampling sites are included in the frame or only coastal households are included in the telephone sampling frame.
- Nonresponse error occurs when some sampled units do not provide data, either because they are not located (e.g., not at home in telephone survey), or they refuse to participate (e.g., will not allow counting or measurement of fish in intercept survey).

Errors of measurement in fisheries surveys include the following:

- Respondent error occurs when the respondent cannot or will not supply accurate information. Examples of this in fisheries survey include prestige bias (i.e., the tendency for respondents to answer in a way that makes them feel or look better) and recall bias (i.e., the inability to accurately remember previous events).
- Interviewer error occurs when interviewers introduce error into the data they collect. For example, such an error would occur if

Box 1.2
Essential Statistical Properties

Bias: The deviation of the expected value (mean) of a statistical estimator from the quantity it estimates. Bias can arise if some units in the population have no chance of entering the sample (e.g., anglers fishing from private land who are not encountered in the intercept survey.) Bias also can arise from inaccurate measurements made on those who do enter the sample, such as incorrect recall by respondents. Bias usually cannot be assessed from the sample itself. If it can be assessed, it usually requires a special study outside the sample.
Variance: The expected value of the squared difference between a statistical estimator and its mean, reflecting the estimator's tendency to differ from sample to sample. An estimator with low variance is said to be **precise**. Variance usually can be assessed accurately from the sample itself. Estimators typically are reported with an associated variance that has been computed from the sample.
Mean Square Error (MSE): The expected value of the squared difference between a statistical estimator and the quantity it estimates. MSE can also be computed as bias squared plus variance. Note that an estimator can be accurate (low bias) but not precise (high variance) or vice versa, but an estimator with low MSE is both accurate and precise. In this report, "sources of error" refers to MSE: bias, variance, or both.
Accuracy: An accurate estimator has a small MSE. This implies that it has little or no bias and a small variance. Sometimes the term "accuracy" is used narrowly to imply low bias only. Note that highly inaccurate estimators are not useful, regardless of their precision. For example, an estimator of harvest that ignores the data and is given as -5 fish every year is perfectly precise (zero variance) but very inaccurate.

the data collectors incorrectly identified a fish species in the intercept sample.

Other issues that need to be addressed in assessing estimates made from a sampling design are whether the sampling and measurement are being conducted as designed and whether the estimation procedure is matched appropriately to the sample design. For example, are the inter-viewers in the intercept surveys being trained and monitored so that they collect data from a probability sample of anglers? Are the data collected appropriately weighted and combined to take into account the complexities of the probability design?

EXISTING SURVEYS

There are several different survey methods currently being employed throughout the United States to monitor marine recreational fishery catch and effort. Primarily, these efforts are conducted and funded by NMFS, but there are many federal–state cooperative programs as well. While the MRFSS has been the predominant survey nationally, some states have never been a part of this program. For example, as alternatives to the MRFSS, Texas has conducted the Texas Marine Recreational Fishing Survey since 1974, and Alaska conducts the Alaska Sport Fish Statewide Harvest Survey. These states generally compile recreational fishing data for summary reports, but the data are not submitted to the MRFSS for inclusion in the national database. And while the MRFSS is still a major survey for some regions, fisheries managers in each region (e.g., Atlantic, Gulf, and Pacific coasts; Hawaii; Alaska) all use a different combination of methods.

Currently, there are at least 13 component surveys conducted by federal or state agencies (Figure 1.2 and Appendix B), largely funded through the MRFSS, that produce data that are compatible with the overall MRFSS goals but that may have significantly different methodologies and statistical properties. The number of surveys has grown as more or better data are needed for a particular sector or geographical area. For example, all states along the west coast have implemented individual surveys that replace the national MRFSS program due to their needs for better assessment data, such as for managing groundfish on an in-season basis. Generally, the MRFSS has worked with partner agencies to develop these supplemental surveys. Because data must be comparable across all surveys for national estimates, NMFS works with the states and the regional fisheries commissions to plan new surveys, test new designs, benchmark data, and analyze and manage collected data.

As mentioned previously, most recreational surveys are designed with two separate components that are then combined to estimate total catch. For the MRFSS, these two complementary surveys are the access-point angler intercept survey used to determine catch rate and species composition and the coastal household telephone survey used to determine fishing effort. The Puget Sound Sampling Program in Washington, the Shore and Estuary Boat Survey in Oregon, and the California Recreational Fisheries Survey (CRFS) are also telephone–access surveys; however, the details of survey design are quite different. Alternately, the

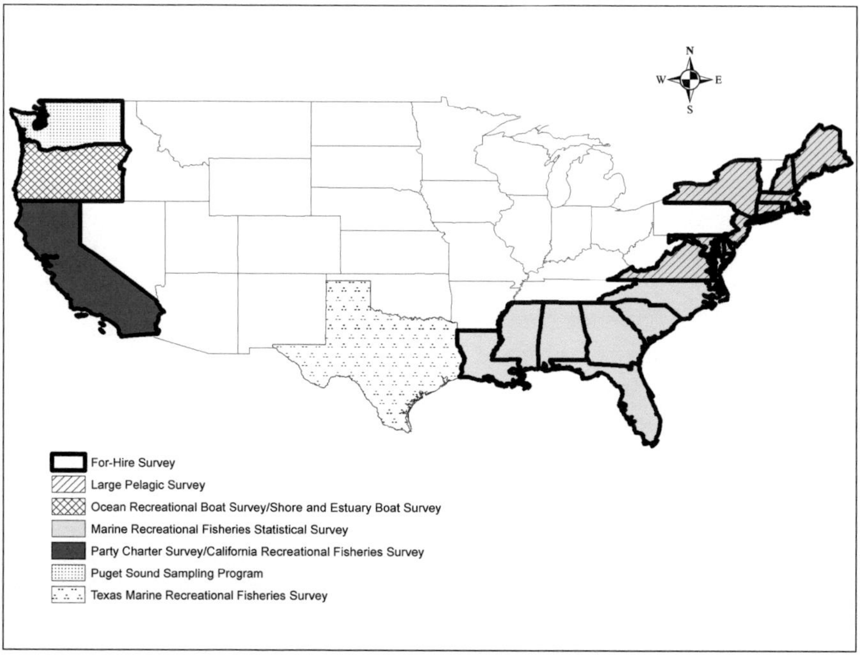

FIGURE 1.2 Many different surveys nationwide are used to estimate recreational catch and catch per unit effort (CPUE) data. The MRFSS is used along the eastern seaboard and the Gulf coast, but this survey has been replaced in each of the west coast states with surveys that are more tailored to state management needs. The MRFSS includes an access-point (onsite) intercept angler survey, a telephone survey (offsite), and the For-Hire Survey, which includes on- and offsite components, and occasionally, observers. An overlap of surveys exists in several states and regions. For example, the For-Hire Survey and the Party Charter Survey are both used in California to measure the for-hire sector. The Large Pelagic Survey is administered along the mid- and north Atlantic coast, as well as the For-Hire Survey and the MRFSS. (Since this map does not show all existing surveys, a more complete list is available in Appendix B.)

Oregon Recreational Boat Survey and the Ocean Sampling Program (Washington) can be categorized as an access–access survey, as are the private boat and artificial structure mode sampling components of CRFS, where boat counts at a specific location are used to determine effort and the access-point intercept is used to gather catch rate and species information.

Due to the diversity of fishing modes and the nature and motivations of the angler population participating in each of these modes, no single survey methodology is likely capable of adequately capturing all the needed data. Recognizing these issues, both the MRFSS and state agencies have augmented the telephone–access surveys with more focused surveys to deal with these other fishing modes. For example, the MRFSS For-Hire Survey was designed in 1995 and has now been implemented nationwide (except for Texas, Alaska, Oregon, Washington, and Hawaii) to provide adequate coverage of the guided trip, head boat, and charter boat sectors. Some species of interest, such as salmon, halibut, bluefin tuna, and marlin, require catch card reporting, which results in catch information for those species, depending in part on how many cards are returned. Further, other rare event species that were typically undersampled in the MRFSS are now targeted with the Large Pelagic Survey.

NMFS, the regional fisheries commissions, and the states have made many advances that have improved recreational fishing surveys in recent years. However, many of the improvements have been made piecemeal as issues or demands arise. Because the context for marine recreational fisheries management has changed for the reasons noted earlier and because these other advances have occurred and other survey approaches have been developed, it is appropriate now to critique the entire concept of a MRFSS or MRFSS-like effort. The goal is to provide recommendations, developed in the following chapters, to facilitate future progress.

COMMITTEE APPROACH AND REPORT ORGANIZATION

To address questions about the MRFSS and other surveys, NMFS asked the National Academies to assemble a committee to (1) review current survey methods used by existing federal and state cooperative programs to estimate marine recreational fisheries effort, CPUE, and catch and (2) make recommendations for possible improvements to current surveys and possible implementation of alternative approaches. (See Box S.1 for the committee's full statement of task.)

The Committee on the Review of Recreational Fisheries Survey Methods, composed of experts in survey design and statistics, biological statistics, fisheries management, and recreational fisheries economics and sociology, met five times over the course of the study (March 10–11,

2005, in Washington, DC; May 19–21, 2005, in San Francisco, CA; July 7–9, 2005, in New Orleans, LA; September 22–24, 2005, in New York, NY; and October 26–28, 2005, in Tampa, FL). At each meeting, there were one or more public sessions during which the committee heard from federal and state government officials (including NMFS officials), anglers, representatives of fishing organizations, representatives of other nongovernmental organizations, and members of the public. (These presenters are listed in the acknowledgments section in the front of this report.) The presentations and written information provided at those meetings and at other times, along with various sources of published and unpublished literature, were taken into account as the committee arrived at its findings and recommendations.

The committee took the approach of discussing some of the major design issues or data needs associated primarily with the MRFSS but that also occur in some of the other regional or state surveys. The report structure is defined by these issues and not by the surveys themselves. The report focuses on marine (and not freshwater) recreational fishing because that is the interest and purview of NMFS.

This report presents general issues and problems with current survey methods and designs in Chapter 2, and Chapter 3 reviews sample design concerns and presents alternative survey methodologies that can be used to improve recreational fisheries data. Subsequent chapters discuss how recreational data are used in stock assessments and some of the issues that need to be addressed if these data continue to be an essential information source for future assessments (Chapter 4) and how new survey methods are needed to collect better quality or more useful data on the human dimensions of recreational fisheries (Chapter 5). Chapter 6 discusses methods for establishing a program structure that may allow for continued improvements in recreational fishing survey methods. The report concludes with a discussion of methods to improve communication and enhance confidence in the national and state survey programs (Chapter 7). Additionally, case studies of three recreational species and their associated survey and management methods are used throughout the report to highlight specific issues; the full details of these examples can be found in Appendix C.

2

Current Situation and Problems in Effort and Catch Estimation

This chapter highlights the complex nature of monitoring fishing effort and catches within the recreational fishing sector, discusses the data collection and estimation challenges posed by this complexity, and focuses on issues associated with the implementation of existing surveys. At present, there is a patchwork of methods and systems of data collection for recreational fishing throughout the United States, primarily as a result of historical anomalies and different regional and state management approaches. However, basic similarities in the methods used by component programs do exist because a two-phased process[1] is generally needed to arrive at an estimate of the essential parameter, total catch, based on information about effort and catch per unit effort (CPUE). Survey programs must also consider design characteristics needed to address the requirements for information on indices of relative population abundance, biological sampling of fish species, and related parameters concerning economics and angler attitudes.

The common feature of catch estimation by surveys discussed in this report is that estimates of total catch for each subcomponent (i.e., the design-based spatial and temporal strata, or the post-data collection strata, defined by species, primary fishing area, and type of catch) are obtained by multiplying together the estimates of effort and CPUE gathered from two separate surveys. Total catch is estimated in this way

---

[1] Note that the committee is not referring to a nested survey process here but instead is using two-phase to indicate the use of two different surveys, one to estimate CPUE and the other to estimate effort.

because of concern that nonsampling errors would invalidate direct estimates of total catch derived from data of either a dockside or telephone survey alone. (See Figure 2.1 for sources of error in survey estimates.) While a telephone survey might theoretically provide access to all anglers, anglers contacted in this manner may provide poor estimates of catch because they would be required to identify species caught or to recall the size or number of fish landed for fishing trips that may have occurred weeks or even months before being contacted. Similarly, reliance on dock-side intercept surveys alone is susceptible to problems of incomplete spatial sampling frames (see Box 2.1 for a discussion about sampling frames) wherein undercoverage bias results from the difficulties of accessing private fishing sites. In addition, poor precision of intercept surveys can result from financial constraints on the number of interviews that can be conducted, particularly to reach sites that are more remote or to sample dispersed but low-use sites adequately. Therefore, the Marine Recreational Fisheries Statistics Survey (MRFSS) uses a hybrid approach in which dockside intercept surveys are used to estimate CPUE and conduct biological sampling, and catch and telephone interviews are used to estimate effort. The results of the two surveys are combined to yield an estimate of total catch. The result of using these complementary strategies for assessing effort and CPUE and for obtaining biological information is that the estimation procedure is more complex than for many other demographic surveys since it requires two separate sampling operations. Additionally, numerous adjustments and extrapolations arise because the sample frames on which the surveys are based are incomplete or unrepresentative of the entire population.

Evidence throughout this chapter will show the fundamental problems associated with the overall national MRFSS program and with some of the component state surveys. These problems are variations on several common elements. There are potentially large biases in the sample estimates, and neither their magnitude nor impact can be measured using the current data. These biases are due to the following reasons:

- The sample frames for both catch rate estimation and for effort estimation are incomplete, contain errors, or both.
- Fidelity to sampling protocols used in both effort estimation interviews and access-point intercept surveys is not monitored adequately.

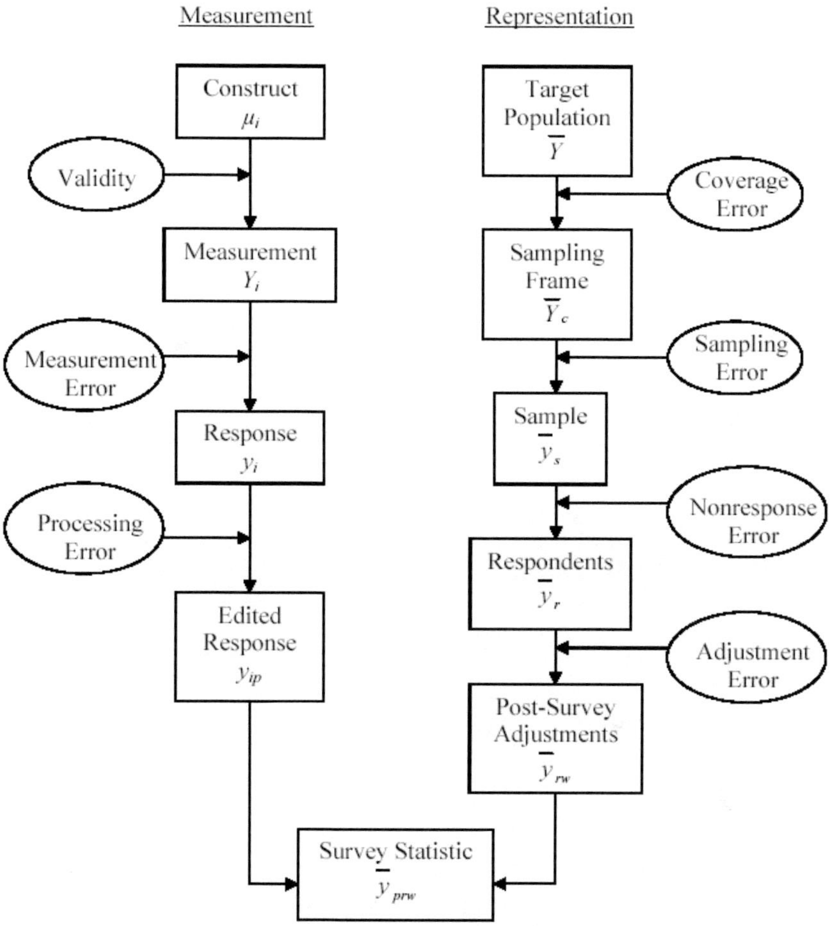

FIGURE 2.1 Sources of error in survey estimates (Groves et al., 2004; reprinted with permission from John Wiley & Sons, Inc.).

- Assumptions of unknown validity are used in the expansion of estimates over the nonsampled segments of the angler population.

Other potential biases within the sampling design can be estimated using the existing data, but these analyses have not been conducted. Inefficiencies arising from overcoverage in the list frame for effort estimation result in low precision of estimates and higher cost than

Box 2.1
Area Versus List Frames and Their Use in Angler Surveys

Frame: A sampling frame is a collection of units from which a sample will be drawn. The frame is ideally identical to the population (a complete frame) about which one wishes to learn, but typically, the frame is a subset of the population (an incomplete frame). If the frame is different from the population in any way, bias can be introduced if the value of a parameter for the frame is not the same as the value of that parameter for the population. Two standard frame types are *list frames* and *area frames.* Coverage errors arise from errors in elements of the frame, more commonly in list frames, and will lead to bias in estimates based on sampling of the frame. *Overcoverage* can arise when frame references exist but do not provide access to sample elements (e.g., licenses without addresses, incorrect telephone numbers, households with telephones but no anglers). *Undercoverage* arises when some population units exist but are not linked to the frame and therefore have no probability of being sampled (e.g., fishing licenses sold that are not recorded in the list frame).

List frame: A list of information that provides direct access to sample units. Through its random digit dialing (RDD) sample, the MRFSS uses a list frame of all working landline telephone numbers in coastal counties. This frame suffers from overcoverage since not all households contain anglers, undercoverage since some anglers do not live in coastal counties or live in coastal counties but have no landline telephones, and duplications since some anglers live in households with more than one working landline. Similarly, the For-Hire Survey uses a list frame of charter boat operators or licenses that may be incomplete. The access-point intercept survey used within the MRFSS and component programs is also an incomplete list frame. Even though the intercept sample sites may be geo-referenced, they are chosen from a master list of documented access sites (e.g., boat ramps, docks, piers) and therefore are not an area frame. Typically, the access site frame will not list all sites, resulting in undercoverage.

Area frame: In the context of site access, an area frame would be a coastline map that could be sampled in portions, and each portion would be searched for access sites. An area frame provides indirect access to sampling sites; access is indirect because the geographic areas must be selected first and the direct access to sample units achieved through a second-stage sampling process. Currently, area frames are not used in the MRFSS.

would be required if the list frame coincided with the angler population. Moreover, the data needs for management and analysis have changed since the inception of the program, including the following:

- Management decisions require data on finer temporal and spatial scales.
- Recreational fishing data are now required for use in stock assessments, sometimes as the sole data concerning stock status.
- Managing recreational catch and retention has become a primary activity for fisheries management as recreational removals have supplanted commercial removals for many species and areas.

Finally, the expertise and personnel needed to evaluate and improve the survey design and execution continually are lacking, and methods used to collect and analyze recreational fisheries data have not incorporated evolving statistical methodology or new innovations and technologies that would improve statistical efficiency and reduce costs.

A number of regional surveys have been developed in recent years with the aim of addressing some of these problems. However, with such a wide range of surveys conducted, it is beyond the committee's ability to analyze all of their individual problems and potential solutions. Consequently, the issues raised in this chapter tend to focus on the MRFSS and the For-Hire Survey, but in most instances, these same issues are also common to the regional surveys. The issues and characteristics described here are not intended to be inclusive; rather they are meant to illustrate the general nature of the sampling situations and resultant problems.

BIAS AND PRECISION

As with all surveys, minimizing bias and maximizing precision of estimators of important parameters are the goals of the recreational fishing survey program. The problem with achieving these goals is that the nature of recreational fishing does not allow for data to be collected for all anglers. Ideally, representative samples that allow unbiased estimation of the catch by the total angler population should be collected. However, resource limitations, survey design characteristics, sample frame errors, and restricted access to anglers in some modes may result in nonrepresentative sampling of the angler population. Therefore, adjustments, or expansions, in the estimation process are employed to account for the lack of information for some anglers. These adjustments require assumptions about the behavior of the unobserved anglers that are of unknown validity. Furthermore, the data do not exist to test the validity of these assumptions or to determine whether they result in large biases. Not knowing whether the adjustments introduce bias, and not being able to test for this bias, creates uncertainty about the quality of the estimates.

Variation in an estimate among years is a source of major debate for recreational fishing surveys—especially where fluctuations in estimates result in equivalent fluctuations in regulations for subsequent years. It may be the case that these fluctuations are real, but they also may be artificial. They may result from low precision in the estimate (which can be corrected by increasing the sample size or sampling efficiency) so that the estimate may be unbiased but may vary from the true parameter value in any given period because of expected variation. It is currently difficult to assess if this is the problem because standard errors may be estimated incorrectly.

Recreational fishing provides formidable challenges in estimating catch, effort, and economic expenditures by anglers, either regionally or nationally, due to the diversity of fishing sites and modes available to anglers. Recreational fishing can be an individual or group pursuit. It can be based on shore or on water and can be conducted on private boats or through a commercial for-hire vessel. Angler trips can originate from private residences that border fishing waters or involve travel over thousands of miles to a departure site, with additional travel on water to the fishing grounds. Effort can range from only minutes of active fishing for anything caught or for a favorite species to multiple-day trips involving multiple targets; often, trips can cover the entire 24-hour period. Furthermore, the target species for anglers may be varied and may include species entirely allocated to recreational fisheries, as well as those from mixed recreational and commercial fisheries.

The difficulties of covering all fishing modes, access points, and duration of fishing has led to several additional surveys that complement the basic MRFSS approach. Yet, even these additional surveys are unable to measure all essential strata, leading to assumptions about unsampled fishing behavior. Below is a brief description of the different angler modes that highlights survey and estimation procedures that are used and how bias or imprecision may be introduced into estimates of effort, CPUE, and the resulting total catch. The issues discussed are not intended to be exhaustive for all surveys or even for a single survey but are intended to emphasize the issues that are described below.

Private or Independent Fishing

Shore-based

Shore-based fishing refers to fishing directly from the shoreline (e.g., beaches, banks, headlands) or from artificial structures, such as docks, jetties, piers, bridges, breakwaters, and causeways. This is the most difficult sampling environment because of private property issues and because of a virtually unlimited number of small access points. Anglers who participate in fisheries from public or commercial property can be intercepted by onsite samplers and can be included in CPUE estimation; however, the extensive amount of publicly available property and structures makes attaining an efficient probability-based sample challenging. Also, some shore-based anglers are not accessible through the public access-point frame used for estimating CPUE because they fish from private property. An angler fishing from a private residence might never be subject to an intercept interview, and therefore, his or her data never could contribute to CPUE estimate. Instead, his or her CPUE would be assumed to be the same as for anglers fishing and sampled though other modes. However, in order to expand the estimates based on sampled anglers to this unsampled portion, the assumption must be made that the species composition and catch rates of these anglers is the same as for the sampled anglers. This is assumed to be true, but data to test this assumption have not been collected. These anglers can be included in estimation of effort through the telephone frame. However, a consistent definition or duration of "angler trip" between shore-based and waterborne fishing is elusive.

Effort for this shore-based private fishing is measured through the MRFSS random digit dialing (RDD) survey, but only for anglers who live in coastal counties. Anglers who reside beyond this area, but who fish from shore in the survey area, are excluded from the sampling frame.

Residential Boat Ramps and Docks

Similar to shore-based anglers fishing from private property, waterborne anglers who launch from private residential property are not normally subject to access-point intercept sampling because samplers do not have access to private residential property. If CPUE for these anglers is the same as for those launching from public access sites, then no bias is introduced from this undercoverage. It seems possible, however, that the experience and knowledge of the local area among anglers in this mode may cause the two groups to differ in CPUE. The effort within coastal households for this mode can be estimated through the MRFSS RDD survey.

Publicly Owned and Commercially Available Boat Ramps and Moorage

This mode is similar to the use of public structures for shore-based anglers in that use of public facilities for boat launching or moorage provides the opportunity to conduct intercept sampling of waterborne anglers. However, sampling this subpopulation of anglers still can be problematic if there are a great many launching sites. The large number of sites and the limited survey budgets and time may result in a tendency to exclude many small sites from the list of sites chosen for sampler coverage. There may also be issues associated with the timing of angler presence at these sites; the intercept sample design must account for any such temporal stratification. Effort for local anglers (those residing in the selected RDD calling area) using this mode will be estimated through the MRFSS RDD survey, but effort for nonlocal anglers will not.

For-Hire Fishing

When anglers go with a guide, charter fishing on boats with crew, or on head boat trips, their participation and removals are estimated through a different framework than that used for private anglers. However, anglers who rent boats for independent, nonguided fishing are captured by the current MRFSS sampling approaches; these waterborne anglers are treated similarly to the private boat anglers discussed above.

Head boats, charters, and guided boats are commercial enterprises, require registration, can be listed, and thus constitute a smaller and more efficient list sampling frame than that of the population of independent anglers. (Only some states have lists based on saltwater fishing licenses.) Effort in the For-Hire Survey, which measures number of boat trips, number of anglers, and areas fished, is determined from boat directory telephone surveys instead of RDD employed in the MRFSS. Because the list frame is complete, assuming that the directory is kept up to date, the potential bias associated with not collecting effort data from noncoastal county anglers is not an issue as it is with the MRFSS. Catch rate, however, may still be collected though dockside interviews, which share the same sampling issues associated with this type of sampling (discussed later in the chapter). In addition to these general problems, there are specific issues associated with the dockside interview for head boats. Each angler's data are likely to be highly correlated. This results in cluster effects that, if not accounted for, can have a significant impact on both the bias and the standard error calculation for the final estimates (see Chapter 3). (Cluster effects also should be expected for nonguided boat anglers, although probably to a lesser degree than for head boats.) In addition, biological sampling of these catches should account for cluster effects, and stock assessment analysts using these data also must be aware of these potential effects.

The for-hire sector can provide an additional unique opportunity for recreational catch and effort sampling because records of angler participation generally are kept by for-hire companies. These records provide two capabilities: direct estimation of fishing effort (and, frequently, catch) and a source of validation for estimates obtained through alternate sampling methods, such as remote-access sampling of anglers based on a different sampling frame. Records of client participation are kept to varying levels of resolution. In the case of guide boats, records normally are associated with individual anglers. For example, guide boats taking anglers for high-prestige species, like tarpon or bonefish, may involve considerable expenditures, and records for an individual angler might have historical and future value for the guide. For head boat and charter boat fishing, records of fishing effort by anglers may or may not be accompanied by removal data at the individual level.

Validation of charter boat records is recognized as an important component and source of error information for the estimation process. Access-point intercept samplers have noted inconsistencies between charter boat logbook records and observed presence and absence information on vessels at their normal home port. It is important to create a rigorous and objective sampling protocol for validations of this type.

Similarly, validation of angler participation also must accompany the use of charter boat data. It is important that charter boat anglers be included in alternate estimations of fishing effort, such as remote-access sampling, so that a validation of charter boat records can be achieved. It also is important, of course, that care be taken not to count anglers twice (i.e., once in contacting them individually and once through the for-hire survey).

The implementation of the Party Charter Survey (For-Hire Survey) in California has improved the estimates of effort and therefore catch by this sector. The ability to define the sampling list frame through a directory of commercial enterprises also has improved the efficiency of sampling these anglers over what had been achieved previously in the MRFSS. In addition, more timely data are provided because a percentage of the vessels within the directory are sampled each week instead of waiting two months, as with the MRFSS. Additional improvements that can be made for this sector are discussed in Chapter 3.

Tournaments are special cases that might have some potential use for assessing biases and for providing information for some species. Although angler catch and effort often are well documented, they do not represent typical angler activities and often focus on highly migratory species, which often are not included in the MRFSS.

Night Fishing

In some areas, night fishing is common and creates unique challenges to estimation of catch rates and, to a lesser degree, fishing effort. Effort for night fishing can be estimated through the telephone survey in the same way as for other modes of fishing. However, estimation of catch rate for this mode is highly problematic because, while anglers participating in this mode may be accessed, in theory, through an existing frame, they are inaccessible because samplers normally do not intercept anglers at night. Therefore, a secondary temporal stratification within the access site sampling frame is required to estimate catch rate by this fishing mode. Such a program has been implemented in the Mississippi Shore Night Fishing Survey. Another method of obtaining angler-supplied night-catch information is to add some questions to the telephone survey; although, this will create additional complexities. In many cases, night-fishing catch will have to be ignored.

Spatial and Temporal Issues of Sampling Coverage

In most cases, CPUE is achieved via intercept sampling at access points. Access points are given different probabilities of selection into the sample, with sites weighted and chosen based on expected angler activity. The set of selection probabilities are referred to as the pressure matrix. This sample design is selected to improve the efficiency of the CPUE estimate and seems likely to do so if the pressure matrix effort estimates are accurate. While the MRFSS and its derivatives have attempted to keep the pressure matrix relevant to current effort distribution, the methods used to update the pressure matrix are not consistent across regions. In addition, the selection probabilities are not used in the estimation process, which will lead to bias in the estimators, except in unusual circumstances. This is discussed in more detail in a later section.

A source of potential bias in the estimate of effort is due to the proportion of private anglers who are not part of the sampling frame. Effort estimation is based on telephone sampling of residents of coastal counties, and many private anglers do not reside in these counties. An adjustment based on information obtained in the intercept sample is attempted, but this will be adequate only under special circumstances. (Again, this is discussed in more detail in a later section.) This mismatch of the frames for estimating catch rate and effort results in a decreased capability for validation of fishing effort estimation through comparison of estimates from the two frames.

The temporal stratification of the current MRFSS is based on two-month sampling periods, or waves. However, the timeliness of the estimation from each wave varies by region. In most regions, the lack of timeliness is not important because species' harvests are not managed in-season. However, for several major species on both coasts and in the Gulf of Mexico, in-season estimation is a key component of management. The timeframe for estimation through the MRFSS process (because it takes a long time to accumulate enough fishing households to have an adequate sample size) does not address management requirements consistently, in part due to the inefficient telephone sampling frame for estimating fishing effort.

GENERAL PROCEDURAL AND ESTIMATION ISSUES

As a precursor to discussing the specific issues of precision, bias, and efficiency, it is necessary to describe the framework by which the estimated total catch is derived. Total estimated catch ($\hat{C}$) (here total catch represents landed and released catch) is the product of total estimated effort ($\hat{E}$), made from the telephone survey, and estimated catch rate of anglers ($CP\hat{U}E$) from the intercept survey. That is:

$$\hat{C} = \hat{E} \times CP\hat{U}E$$

Effort, in turn, is estimated as the ratio of the estimated total number of trips for those households sampled in the RDD frame ($\hat{E}_{RDD}$) and the estimated proportion of anglers who are sampled using the RDD frame ($\hat{\pi}_{RDD}$) (this number will be less than 1 since some anglers reside outside the frame, in noncoastal counties, and is estimated from the access survey):

$$\hat{E} = \hat{E}_{RDD} / \hat{\pi}_{RDD}$$

In other words, total estimated effort is the corrected effort after adjusting for the proportion included in the sampling frame $\hat{\pi}_{RDD}$. The variance of total catch is estimated using a delta method variance estimator for products and ratios. These estimates for individual areas or strata are summed across areas or strata to obtain aggregated estimates.

Accounting for Anglers Not Included in the Effort Frame

As previously discussed, some anglers are not surveyed for effort because they reside outside the coastal county RDD sampling frame; therefore, the frame used for estimating effort is incomplete. That is, the sample frame includes only a subset of the true population (i.e., those in telephone households in coastal counties), and estimates derived from sampling of this frame are then adjusted upward by "expanding" the frame. The expansion factor is generated through the intercept survey; when interviewed on the dock, anglers are asked where they live. The proportion of RDD-in-frame to total anglers (both in- and out-of-frame) is used to estimate π_{RDD}, and subsequently to expand $\hat{E}_{RDD}$ to total estimated effort.

In theory, this is a reasonable expansion, but in practice, there are some concerns with this method. In particular, the intercept frame that is used to correct for the incompleteness of the effort frame is itself incomplete. Private property and docks (without public access) are not included as sampling sites. Therefore, the estimate of π_{RDD} likely would be biased downward since noncoastal residents might be most likely to fish from areas that are subject to the intercept surveys. Conversely, if these anglers do not live on the water, they most likely will access the water through public sites (although a component may visit friends or relatives with private access). Therefore, a smaller fraction of coastal county residents most likely would be intercepted than there are in the angler population. If this sample proportion is biased downward, it will result in an overestimate of effort. However, the key point is that there is no information in the data collected that allows analysis of this bias. In addition, the intercept data may be biased simply due to a lack of a true probability sample because interviewers on the docks in many cases are allowed considerable freedom in choosing the sample and may select interview sites based on sufficient numbers of anglers, their receptivity, or other factors. This issue is discussed in more detail in a later section.

Inefficiencies in the Effort Estimation

RDD, even limited to coastal county residences, is not the most efficient way to gather angler effort information. In urban areas, less than 1 in 20 of the telephone intercepts reaches an angler. Improving the process whereby anglers are identified and contacted would not only improve the quality of the estimates but also should reduce costs. Remedies exist for other inefficiencies as well. For example, under the current sampling regime, identifying an angler costs more than the taking of information once the angler has been identified. Under this high ratio of costs for identifying anglers to costs for information gathering, it may be more efficient to repeatedly sample identified anglers (i.e., a panel survey). This also would allow comparison of angler effort levels and perhaps catch rates over time, yielding more precise estimates. Even if the costs were comparable, repeated sampling may be preferable because accuracy in trend detection may be improved.

Accounting for Anglers Not Included in the Catch per Unit Effort Frame

Estimation of CPUE is achieved via intercept sampling at access points. However, the list frame used for the intercept sampling program is incomplete. The sample cannot yield information about CPUE from anglers who fish from small public access sites not included on the frame, anglers fishing from private shorelines, or those who leave from private docks. For the sample estimate of CPUE to be unbiased, one must assume that anglers who fish at sites not included on the intercept sampling frame have the same success in fishing and that they target the same species as those who fish at sites that are included on the frame. It also must be assumed that the unit of effort, the trip, is equivalent in the two groups. In fact, it seems likely that a trip actually has a different meaning for those who fish from their own private dock and those who do not. If access is easy, such as having a dock or boat near an angler's home, a "trip" could last only a couple of hours. Alternatively, for an angler who is paying for a charter, fishing may last the entire day. The catch per trip for these two groups could differ only due to the length of the trip. There is no way to assess the bias due to this undercoverage since no information about catch rate is available for anglers at the sites not in the frame.

The daily timing of intercept surveys presents a second source of undercoverage. Most intercept surveys sample only during daylight hours and may have considerable flexibility in timing of sampling. In many venues, anglers do not return to access points until after dark or may engage entirely in night fishing. These anglers therefore are not captured in the CPUE survey frame. This unsampled component has been recognized through the development of some special-purpose surveys, such as the Mississippi Shore Night Fishing Survey, but such surveys are limited in number. Catch rates, by species, from daytime fishing are applied to the night fishing mode in the same fashion as extrapolations to other unobserved fishing, but such assignment of both catch rate and species composition may not be justified and may diminish the accuracy of the estimates. Even if catch rates are the same for night fishing, the angler community recognizes this undercoverage in the intercept frame and may perceive the results to be less credible.

The intercept frame contains a list of angler access sites, along with estimates of fishing effort at those sites. The sample design uses unequal probability of selection of access sites, with selection probabilities proportional to estimated fishing effort. However, these measures of effort are not used in the estimation procedure, at least as carried out by the MRFSS. Instead, the design is treated as though it was self-weighting, and an unweighted estimator is computed. This estimator will be biased, except in very special circumstances.

There are two conditions under which the use of the unweighted estimator would not result in a bias in estimation. If there is no difference in CPUE (by species or type) by anglers using different access points, then there will not be bias resulting from ignoring the weights. Alternatively, a self-weighting estimator would be unbiased if the selection probability is actually proportional to the number of angler trips at each site and a constant number of anglers were sampled at each selected site. In that case, the probability of selection for each angler trip would be:

$$(t_i / t)(k / t_i) = k / t$$

where t_i is the number of angler trips at site i in the period, t is the total angler trips, and k is the number of angler trips selected at each site. If neither of these is true, however, then bias in the estimation of CPUE could result. The validity of this estimate could be investigated from the sample data by comparing estimated CPUE in sites with low and high selection probabilities. However, one could not assume that because this is true for a few subpopulations that it will remain generally true for all time, geography, species, and catch type cells. Rather, one would need to monitor this assumption regularly or develop an estimation system that does not require such an assumption (e.g., using weights computed from the design).

The Possibility of Introducing Bias into the Intercept Survey

It is randomness in the sampling process that assures that a probability sample is achieved. More specifically, in order for a sampling method to be called a probability sample, each unit must have a positive and known probability of selection. Thus, a probabilistic algorithm must be used (either before interviewers are given assignments or by the interviewer in the field) to decide which units will enter the sample and which will not. In a probability sample, interviewers will exercise no judgment in choosing who to interview. To the extent this protocol is followed, the laws of probability will provide estimators with known properties.

Currently, the onsite intercepts for all recreational surveys are assumed to be a random sample. However, the collection of intercept data has been tailored to the kinds of access sites that are present in a particular region, and interviewers frequently are allowed to make judgments about where, when, and which units to sample. This means that these samples may not be true probability (random) samples. Generally, the leeway afforded to onsite samplers is an attempt to reduce costs by reducing the time it takes to gather the target number of samples. This is problematic because such sampling is, in essence, a quota sample,[2] rather than a probability sample in which all anglers have a known probability of being intercepted. This deviation from a probability sampling protocol has an unknown impact on estimates of both CPUE and effort.

Besides the deviations from sampling protocol that are explicitly allowed, there may be other instances in which interviewers stray from instructions on sample selection. There is no regular interviewer monitoring program included in the sampling protocol, as is common in most survey operations. Indeed, it would be difficult to use the most common types of interviewer quality control programs in the intercept survey setting because they are based on a reinterview of a sample of respondents. The result of this problem is that it is not known, nor is there an easy way to determine, how much interviewer error affects the quality of the data gathered through the intercept survey. Making the development of a reinterview program especially difficult is the fact that intercept interviews are conducted by a wide range of people. Several states either have their own intercept surveys or have taken over the conduct of this portion of the MRFSS, but still others rely on contractors to complete the surveys. With multiple organizations involved, it is difficult to specify and monitor adherence to a common sampling protocol across survey efforts.

---

[2] A quota sample defines groups of people who are deemed important to reach, based on information about the target population. Quotas are set for each group based on the group's relative size in the population. Quota samples are not random samples, and their use can lead to bias if anglers who are difficult to reach differ from those who are easy to reach. In addition, the precision of estimates cannot be calculated (Pollock et al., 1994).

Variance Estimation

Standard error is an important indicator of the quality of an estimator, and yet, correctly estimating standard error may be the most difficult part of the estimation process. The sample design and estimator form for the MRFSS are nonstandard, so correct assessment of variance is challenging. There appear to be some problems with variance estimation as it is carried out currently. Addition of variances across subpopulations to obtain valid estimates of variance for aggregates requires that the estimators within each subpopulation be independent. For some subpopulation aggregates, this will be valid. For example, for strata representing time periods, it seems a reasonable assumption since both intercept and telephone samples are drawn independently in different time periods. (However, for sparse subpopulations, information from past time periods are imputed, which would invalidate this method.) For aggregations over post-strata, such as catch type (e.g., removals consisting of catch available for inspection [A] and catch unavailable for inspection because it is filleted, discarded dead, or refused for inspection [B1]), this is unlikely to be valid since the same sampling units are used to obtain each type of data. There is information in the data that would allow this correction to be attempted; in other words, it is possible to calculate correlations from the sample, and the variance estimation method could be changed to account for the correlation.

Estimating Mortality

The issue of catch and release of both target and nontarget species requires much greater attention and estimation of associated mortality rates than has occurred to date. Currently, catch released alive (B2) usually is not incorporated into the catch estimates used for quota monitoring because there is no verification of the catch and it ostensibly is released alive, although Oregon and Washington do apply hooking mortality rates to discards in ocean boat fisheries that vary by species and other factors. However, this assumption ignores the high hooking mortality rate associated with some fisheries, especially for those species with swimbladders that are caught at significant depths, such as rockfish or grouper. In some instances, mortality on released fish may represent the major mortality factor in total removals. For example, several species of west coast rockfish are under severe restrictions of total allowable catch, and the recreational fishery (both retained and released) is the primary source of removals.

Measuring the release of species is problematic because it relies on angler recall (in the absence of onboard observers) and angler knowledge of species. Estimates of size are even more difficult to collect and most likely are overestimated due to prestige bias or are subject to rounding errors. Better methods to estimate the number of released fish are needed. Some of the specific issues related to catch-and-release fisheries include (1) released catch cannot be inspected in an onsite survey, unlike the kept catch; (2) rounding errors are common; (3) exaggeration or under reporting due to memory problems are possible; (4) species identification errors may occur; and (5) the size and age distribution may be different compared to kept fish. All of these errors can be serious. Observers in boat-based recreational fisheries can be used to obtain direct estimates of fish release numbers. However, releases are likely to be different if observers are present.

Research on released fish mortality estimation from cage studies and tagging studies is needed to help estimate the contribution of the auxiliary information collected about the depth from which fish were caught. However, cage experiments would provide only a minimum estimate of mortality and reflect only "physiological" mortality rather than the "ecological" mortality that would be measured through tagging studies.

NEW DEMANDS ON RECREATIONAL FISHING DATA

This committee identified a number of areas in which designers of sampling programs, data collectors, and users of recreational fisheries data appear to have incomplete communication, mismatched criteria, or other miscommunications. In most instances, these issues have arisen because the current uses of recreational fisheries data were not anticipated in the design of the MRFSS. Current users require data with higher resolution—spatially, temporally, and taxonomically—than the current MRFSS design can deliver.

Two of the major recurring issues facing recreational surveys are adequate spatial and temporal resolution. These needs are driven primarily by the type of management applied in each area. Management tactics have changed since the inception of the MRFSS and continue to change as more stocks are monitored and managed. Survey designs now require greater coverage and more detail to estimate harvest and effort for national and interstate management (Henry, 2002; Lyle et al., 2002; Pollock, 2002).

In addition, recreational data are now used for many stock assessments. For stocks that now have low catches from the commercial sector, recreational fishing data may be the primary data on which stock assessments are based. In addition, topographic and other differences between the coasts in various regions also affect demands on sampling design. For example, the rocky coasts of the Pacific Northwest, with their rough seas, provide far fewer potential access points for boats than the sandy coasts and calmer waters of Florida and the Gulf of Mexico. The application of recreational data to finer management scales and their use in assessments have highlighted potential issues with the bias and precision of current survey methods that previously were less important. However, if these uses of the data are to continue, changes in the survey methods are required to provide the needed information.

Spatial and Temporal Resolution in Catch and Effort Estimation

Currently, many fisheries are monitored at the state level, which is a finer stratification than intended originally for the data collected. In order to provide state estimates with reasonable precision, many states have increased their sample size, either by adding additional sampling by state personnel or by asking the MRFSS contractor to complete more calls and onsite intercepts. These actions, taken on the whole, seem to result in more precise estimates of total catch within these smaller areas. In addition, these measures appear to have increased angler confidence because increasing sample size is a straightforward premise that non-survey scientists can understand. In some cases, it also presents the state as taking a proactive approach that is appreciated by anglers—the states are no longer just saying that the data are not good enough to manage, they are actually doing something about it. Of course, additional samples require more money, but if quotas are to be allocated and monitored by each state, these additional samples are necessary.

There are numerous other methods that can be used to increase precision on smaller scales than are employed in other national surveys. However, these methods generally have not been explored for their application to recreational fishing surveys. Some of these methods are discussed in Chapter 3. However, data gathering on smaller scales will only be useful if the data collection methods are not biased and the assumptions made about extrapolations and imputations are valid, as discussed in the previous section.

The temporal scale of data collection also continues to be pushed to a finer level of resolution than originally intended. Within the MRFSS, effort sampling is conducted in two-month waves. Checking the data and completing the final estimates takes another two months, meaning that catch estimates are not released until at least four months following the actual fishing effort. This lack of timeliness raises many issues for managers.

Obviously, this time lag does not allow in-season management, which is why different surveys have been implemented in states wanting to manage in-season, such as California, Washington, and Oregon. These states have reworked the fundamental components of the MRFSS—the intercept survey and the telephone survey—in order to compile more timely data. The most fundamental change is the implementation of an angler registry so that the sampling frame used to determine effort is more defined and efficient than that of RDD of the MRFSS. Finer spatial-scale management also has required larger sample sizes for each sampling wave to ensure sufficient precision of resultant estimates.

Even with annual management, data timeliness is an issue. Often, the data from the previous year have not been analyzed completely until the following season is under way. This can result in adjustments to the total allowable catch once the season has begun. While this is not truly in-season management, the effects can be similar if the adjustments to the current year mean that fewer fish can be taken or if the season has to close earlier than expected. These situations are difficult for anglers and operators of for-hire vessels to deal with. For example, fishing trips can be planned many months to a year in advance; yet, there may be no guarantee the fishery will still be open for future planned trips. This uncertainty is perceived to be a much larger problem in recreational fisheries than in commercial fisheries because anglers can be infrequent users. The time it takes to collect, verify, and calculate fishing effort and catch using conventional survey approaches is too lengthy, even for annual management, if stability in the yearly total allowable catch is desired.

Use of Data for Stock Assessments

A large mismatch appears to exist between recreational sampling programs and the stock assessment scientists using them. The MRFSS and even some of the newer surveys were not designed to gather data for stock assessments; yet, the estimates of total catch and the biological data collected during the intercept survey are often used in assessments.

Assessment scientists using these data generally do not have a clear understanding of the data collection process, or the data collection process may not be executed in the manner assumed by the scientist. The implications of this mismatch between those collecting and those using the data are profound. The lack of continuity in intercept samplers, differences in sampling methods applied to different modes of fishing (e.g., shore-based and boat-based private anglers or those using various for-hire vessels), differences in sample element definition, lack of incorporation of design elements in the estimation process (e.g., weighting of spatial or temporal sampling strata), lack of consistency (or accuracy) in species designation among fishing or sampling modes, and the inability to combine information based on different sampling modes all compromise the inclusion of these data in the assessment process. Data from different sampling modes may have unknown statistical properties because the data collection emerges from the implementation of general designs that are adapted to suit local circumstances. Scientists using these data may assume that their statistical properties are known and estimable. (More specific problems associated with recreational fisheries data and their incorporation into stock assessments [e.g., the difficulty of measuring which, if any, species are being targeted] are discussed in Chapter 4.)

A common knowledge base among anglers, data collectors, and data users is required if surveys are to fulfill current data needs. This is not to say that all anglers must have a complete knowledge base of species, but the intercept samplers and the anglers must categorize catch to a jointly understood level. This is particularly important for taxonomic stratification of data. Stock assessment scientists must be able to employ data with confidence that species designations are applied accurately and consistently in the sampling process or with knowledge that higher groupings of taxonomic categories are used.

INCORPORATING NEW IDEAS AND TESTING OLD ONES

Surveys designed for monitoring long-term status of populations have considerable inertia and are resistant to change. In part, this resistance is appropriate if the data provided by the surveys are to be consistent and useful over long periods. Major design changes can break the continuity of data and render them unusable for population monitoring unless the new and old surveys are run in parallel for some years during the change. In addition, the original design objective for a given survey may continue to be relevant, in spite of subsequent objectives that may be of equal or greater priority than the original. Resistance to change also arises because of the fixed commitment of resources (human and material) to existing designs.

The MRFSS has made some changes to accommodate better estimation of some fishing modes (e.g., For-Hire Survey of charter boats). However, the fundamental aspects of the two-phased survey have not evolved significantly since the inception of the program. Different survey designs exist that could possibly improve the quality of the collected data; yet, few new approaches have been undertaken by the MRFSS or the state surveys. Indeed, several previous reviews have offered suggestions for improvements, but most of these, including several from a previous report by the National Research Council (2000), have not been implemented, perhaps due to a lack of staff and additional funding.

While several external reviews of the MRFSS or portions of it have been conducted (Essig and Holliday, 1991; Guthrie et al., 1991), there is presently no internal process of user feedback on evaluation and modification of the design within the MRFSS. Some users of recreational data have initiated dialogue with the survey project managers to address specific design issues, but the need exists for a structural feedback process. The rapid evolution of uses of and needs for data from recreational fisheries underscores the requirement for ongoing evaluation by survey managers.

OUTREACH

The committee heard from numerous groups and individuals expressing a lack of confidence in the estimates produced by several of the recreational surveys. While this is not a problem with the survey methodology per se, increasing understanding and confidence in the programs can be as essential as improving the data itself. The credibility gap arises from several causes, including a belief that alternate data sources are more credible; criticism of the temporal, spatial, group, or taxonomic stratification of the intercept sampling; lack of understanding of statistical methodology; or recognition that the existing sampling frames do not describe the angler population adequately.

In addition to dialogue on design issues, survey managers also need to advise data users on constraints to some uses, as well as on fundamental features of the data collection system. The websites for the MRFSS and the regional data programs (National Oceanic and Atmospheric Administration, 2005b) are information rich and provide general background for the average angler. In addition, in recent years, the MRFSS personnel have begun to conduct regular meetings with users to review results of sampling waves. However, the committee heard of a number of instances where users extracted sections of data histories but were unaware of the data characteristics, the methods of compilation, or the fundamental nature of sampling estimation versus census. These observations indicate that while the program has undertaken some outreach activities with users, misconceptions and lack of clarity on data characteristics continue to exist. Further, the lack of user understanding of the design basis of the survey clearly has created some lack of trust even in the underlying data. Considerably greater outreach effort appears necessary but with the recognition that user distrust may not be overcome completely.

CONCLUSIONS AND RECOMMENDATIONS

The designs, sampling strategies, and collection methods of recreational fishing surveys do not provide adequate data for management and policy decisions. Unknown biases in the estimators from these surveys arise from reliance on unverified assumptions. Unless these assumptions are tested and the degree and direction of bias reliably estimated, the extent to which the biases affect final estimates will remain unknown. The statistical properties associated with data collected through different survey techniques differ and often are unknown. The current estimators of error associated with various survey products are likely to be biased and too low. It is necessary at a minimum to determine how those differences affect survey results that use differing methods. It is impossible to assess the adequacy of recreational fishing surveys, particularly those associated with the MRFSS, when potential biases exist. Identifying and eliminating the sources of bias or estimating and correcting for the degree of bias is a fundamental requirement for the provision of reliable estimates from the MRFSS.

The statistical properties of various sampling, data-collection, and data-analysis methods should be determined. Assumptions

should be examined and verified so that biases can be properly evaluated. The complexity of the recreational fishing surveys makes them susceptible to many forms of bias. Unfortunately, it is difficult to avoid this complexity due to the diverse nature of recreational fishing. However, it is important to eliminate sources of bias or appropriately adjust for them when bias is unavoidable. Biases can be addressed through expanding the sampling frame to better represent the population, through experiments used to derive the appropriate correction factors, and through better training and monitoring programs aimed at improving the quality of data sampling. All of these approaches are discussed in the following chapters.

Some of what is viewed as bias by the public can be the result of variance of the mean estimate, arising from inadequate sample size or other sampling errors. While reduction in variance often can be achieved by increasing sample size, improving the statistical efficiency through appropriate choice of estimators and careful implementation of sampling protocols can also be useful. Improved precision commonly is achieved by increasing sample size, and improvements can be gained for estimates derived from recreational fishing surveys through just such an approach. However, gains in statistical efficiency also may be achieved by considering alternate estimators that make better use of the information available and by identifying and implementing mechanisms that improve the effectiveness of the sampling procedure, as for example through the creation of a complete sampling frame of anglers. Improvement will come not only as a result of greater precision but also in terms of reduced sampling effort and cost.

Greater demands on recreational fishing data from both the science and the management sectors are being made. **Management decisions are often made at finer spatial and temporal scales than they were earlier, the mix of recreational and commercial fishing has changed for many areas and species, and stock-assessment models now make greater use of data from recreational fisheries.** Reallocation of harvest from commercial to recreational sectors has increased the need to gather stock assessment information in greater detail from recreational fisheries sources. As managers use recreational data on finer spatial and temporal scales, issues of precision and bias become more pronounced. Existing spatial and temporal sampling strata may be of too coarse a resolution to generate estimates that are adequate for the management requirements.

The MRFSS is in need of additional financial resources so that technical and practical expertise can be added to assist in a major overhaul of the design, implementation, and analysis of data from

the MRFSS. The goals and objectives of fisheries managers, as well as the different surveys, are evolving constantly. There has been progress in survey programs directed to some targeted fisheries with the implementation of new, tailored surveys; yet, additional improvement is required. There have been several reviews of the national program in the last 10 years, but a more fluid, continuous review and feedback would allow for evolution of the program to meet emerging needs. In addition, as statistical theory and sampling technologies improve, it is essential that the managers of these regional or national monitoring programs have greater access to expertise in statistical analysis and sampling design. It appears that the implementation of new survey methods is hampered by the inertia of existing surveys and that even when a need for change is identified, a lack of resources, staff time, and expertise may prohibit implementation of such changes. Development of new survey methods could be accomplished by an external, independent research group as discussed in Chapter 6.

3

Removal Estimation: Alternative Survey Design and Analysis Method

Angler surveys that are well designed, soundly executed, and carefully analyzed with modern statistical methods are crucial for providing high-quality information on total fisheries-related removals and related parameters (fishing effort) on which to base sound fisheries management decisions. As stated in Chapter 2, and now iterated, the important parameters to estimate from a recreational fishing survey are total recreational fishing effort, total recreational harvest (kept catch), and total recreational released catch. Effort (E) is often estimated from one survey and harvest per unit effort (HPUE) and released catch per unit effort $(CPUE)_{Released}$ from a second survey with total harvest (H = A + B1) estimated as:

$$H = E \times HPUE$$

and total released catch (C_R) as:

$$C_R = E \times (CPUE)_{Released}$$

In addition, the fraction of the released catch that dies needs to be estimated in "hooking" mortality (M_H) studies. This enables the estimation of total recreational fishing removals (R), which consists of the kept catch plus the released catch that dies as:

$$R = H + C_R \times M_H$$

This then becomes the basis of stock assessment models. (See Chapter 1 for the definitions of harvest terms.)

Harvest and total removals need to be measured for species and species complexes, specific spatial regions, and temporal periods, depending on the management needs involved. Further, total removals need to be assigned to age or size classes. Also, due to the very different nature of the for-hire and general fishing sectors, these sectors also have to be sampled separately using different methods for efficient estimation.

SMALL- TO LARGE-SCALE SURVEYS FOR SOUND FISHERIES MANAGEMENT

To estimate primarily angler effort and harvest, angler survey design has received much attention since the 1990s when the American Fisheries Society commissioned a symposium and a detailed monograph on the subject (Guthrie et al., 1991; Pollock et al., 1994). The traditional access and roving surveys developed in the 1960s (Robson, 1960, 1961; Malvestuto, 1983) for small water bodies (e.g., lakes, reservoirs, trout streams) were just not suitable for the larger spatial-scale surveys, which are so crucial in fisheries management. This is especially the case in marine fisheries management where the unit of management may range from coastal waters of a small state, to a region involving groups of states, or even up to the national level.

One traditional survey is the access-point intercept survey. Robson and Jones (1989) developed a modification called the "bus route design" and applied it to a small regional-scale fishery in Lake Ontario tributaries in New York. Related access-point marine surveys at the regional scale are run in Texas and Oregon, among others that came under the mandate of this report (see Appendix B). Unfortunately, there are several problems with using these designs, and without major modification and enhancement, these problems limit the usefulness of these surveys. There may be a large number of access points and some may be very small in size; often there is private access that cannot be sampled using only public access points, and the spatial scale may be so large that cost savings may be achieved by using an offsite contact method (e.g., telephone). Roving surveys using agents on foot or in boats also become impractical when it comes to larger spatial scales.

An active area of research involves the design of complex surveys for even larger regional and national marine fisheries (Dauk and Schwarz, 2001; Lyle et al., 2002; Henry, 2002; Pollock, 2002; Volstad et al., in press). Often these surveys require a design that uses one survey for effort and another survey for catch rate. Examples include the pairing of aerial surveys of effort with access surveys of catch or telephone surveys of effort with roving surveys of catch. These paired surveys are known as complemented surveys (Pollock et al., 1994).

One example of a regional survey that uses an important complemented design (aerial and access) is the Georgia Strait Creel Survey. This survey has been run since 1980 by Fisheries and Oceans Canada for the Georgia Strait area near Vancouver, British Columbia. It uses aerial flights to estimate angler effort by taking aerial counts of boats fishing and expanding these counts. This effort estimate is combined with data collected by clerks stationed at access points to record catch rates of individual anglers to estimate total catch. Catch and effort statistics for this tidal sport fishery are calculated for each month and statistical area, and for individual species. According to survey results, catch of salmon species has shown serious declines since 1980 (Hardie et al., 1998; Dauk and Schwarz, 2001). Surveys with this design also are used by Michigan on many of its Great Lakes Surveys (Lockwood et al., 2001) and also in the Delaware River Creel Survey (Volstad et al., in press). The latter survey was designed to estimate catch for important anadromous species (e.g., shad, striped bass) in Delaware and Pennsylvania.

Unfortunately, in many settings, there is a need for more information at much larger regional and even national scales that will require the abandonment of direct onsite estimation of fishing effort for total cost reasons. This suggests the possible use of telephone–access and telephone–telephone survey designs (Pollock et al., 1994). This committee was formed because of a concern for the reliability of a large spatial-scale telephone–access survey, which is what the Marine Recreational Fisheries Statistics Survey (MRFSS) uses (Essig and Holliday, 1991; see Chapter 2). Another national survey run recently in Australia used a telephone–telephone survey design with anglers contacted repeatedly using a panel diary approach (Henry, 2002; Lyle et al., 2002).

The objectives of the Australian survey were to describe the characteristics of anglers (participation rates, sociodemographics); evaluate effort and catch by species, mode, and region; assess economic impacts in terms of investment and expenditure associated with fishing; and evaluate awareness and attitudes to fishing-related matters. All saltwater and freshwater fishing activities were included within the scope of these surveys, which were comprised of the following components:

1) A screening survey designed to identify fishing households and to invite anglers to participate in the follow-up diary survey
2) The diary survey in which fishing and expenditure activity was monitored over 12 months through regular telephone contact by survey interviewers
3) An attitudinal survey administered as a final telephone interview at the completion of the diary survey

In general, an advantage of the use of telephone surveys is that one can obtain information on effort and catch rates for anglers not easily reachable in an onsite survey (typically an access survey). These could include night anglers and anglers fishing from private docks and jetties. However, a key concern is that effort and catch-rate data that are self-reported may contain large measurement errors. These errors may be due to willful deception, recall bias, prestige bias, or lack of knowledge (e.g., species identifications). Lyle et al. (2002) discuss these potential problems and review the methods that they used to attempt to reduce these errors to a low level. In the Australian context, it was not feasible to go to the telephone–access design for cost reasons. It is widely known that there are tradeoffs between survey costs and the precision of the estimates, but it is also true that methods that reduce bias in the estimates may be much more expensive. Onsite catch-rate estimates are much more expensive than offsite self-reported catch-rate estimates (Pollock, 2002). An access survey for catch rate would get around these problems (Essig and Holliday, 1991), and this was an important reason for the current MRFSS design.

What are some appropriate combinations of contact methods to use in particular situations? The spatial scale of fisheries management decisions will be a crucial component. For some local or regional fisheries, the access–access surveys may be optimal; whereas, for other regional surveys, the aerial–access design may be preferred, and at larger scales, the telephone–access (augmented with special studies) is often the only practical option for both the general angler and the for-hire sector. Telephone–telephone surveys, while useful in Australia, will not be useful in the U.S. marine setting to estimate removals for management decisions, as there is the need for an onsite interview component in all surveys. However, telephone–telephone surveys may be useful in special studies of night and private-access fishing because these modes cannot be well assessed in the MRFSS (Chapter 2). General questions that involve policy and economics also could employ telephone panel surveys

(Chapter 5). Augmentation of telephone contacts by internet surveys needs to be considered and will be discussed later in this chapter.

ANGLER SURVEY FRAMES

For estimation of removals and related parameters (effort and CPUE) for marine recreational fisheries, frame problems are extremely challenging. A frame is a set of units that are somehow linked to the population elements of interest. Estimation of a population characteristic is carried out by sampling units from the frame, identifying the population elements linked to the sampled units, and measuring the variable(s) of interest on the population elements. Two standard types of frames, also discussed in Chapter 2, are list frames and area frames.

A list frame with known undercoverage, but that is inexpensive to sample, may be combined with an area frame or another complete list frame that is expensive to sample. Such surveys are called dual-frame surveys. To illustrate, consider the simplest dual-frame estimator, called the screening estimator (Hartley, 1962). The general idea is that the list frame is incomplete; whereas, the area frame is complete, and therefore, there are two components. The overlap domain ($\hat{Y}_{OL}$) is the list frame, and the nonoverlap domain ($\hat{Y}_{NOL}$) consists of those members of the area frame that are not on the list frame. Therefore, assuming simple random sampling in each frame, an estimate of the population total ($\hat{Y}$) would be the sum of the population estimates for the two domains:

$$\hat{Y} = \hat{Y}_{OL} + \hat{Y}_{NOL}$$

where the usual population total estimator for the list frame is used for the overlap domain. All units that are on the list frame are screened out from the area frame, and only the remaining units are used in a standard estimator to get the nonoverlap domain estimator. There are many complications when dual frames are used in real surveys, but this illustrates the general principles. In some of the applications of most interest here, the complete frame would be a random digit dialing (RDD) telephone frame (instead of an area frame), and the incomplete list frame would be a telephone list frame from an angler license file that suffers from incompleteness.

Possible Frames for Effort Estimation

The population characteristic of interest for effort estimates is total angler effort (e.g., number of angler days, number of angler trips). While the description of the unit of effort might vary somewhat among angler modes, the following discussion uses angler days as a surrogate for all of these units of effort. Angler effort can be assessed by either defining the population as all fishing days and then counting anglers active on those days or by defining the population as all anglers and then counting the days they fished.

The first option is problematic because there is usually no simple way to count active anglers on a given day. There are a huge number of ways in which anglers can access the water, though this varies greatly from region to region. Sampling from area frames of coasts and coastal waters could be very inefficient, except in certain constrained waters (e.g., bays, estuaries) in which fishing effort could be assessed through aerial surveys, as mentioned earlier in this chapter, or other direct observations (e.g., bar crossings from the Columbia River in the Oregon Recreational Boat Survey; see Appendix B). Sampling from access-site list frames is used in some smaller regional surveys to get at effort through on-the-ground assessments, such as counting boat trailers or empty marina slips. However, many surveys cover such a large spatial area that this becomes completely impractical. Other difficulties with using access-site list frames are discussed further when considering CPUE estimation through angler intercepts.

The second option of sampling the population of *all* marine anglers is currently problematic but offers the best hope for sound future surveys. It depends on the availability of a list (frame) of the population of all marine anglers. Such license file lists are available in some states but not others; in general, states in the northeastern United States (New Jersey northward) do not have saltwater licenses at all. At the inception of the MRFSS, license frames were not available in many states and the MRFSS designers were forced to use a different list frame. Through its RDD sample, the MRFSS uses a frame of all working landline telephone numbers in coastal counties. This frame suffers from overcoverage since not all households contain anglers, undercoverage since some anglers do not live in coastal counties or they live in coastal counties but do not have landline telephones (a problem likely to grow as more households move to only cellular telephones), and duplications since some anglers live in households with more than one working landline. Overcoverage leads to severe inefficiency in the RDD sampling effort. Undercoverage in the coastal county frame may lead to serious bias since anglers from noncoastal counties are likely to have different effort characteristics than those from coastal counties. An attempt has been made to adjust for this potential bias using information collected via field intercepts in a procedure much like a dual-frame survey; however, as mentioned in Chapter 2, this procedure is ad hoc and likely biased.

Other list frames used in sampling the population of marine anglers include state- or regional-level licensing systems (Washington, Oregon, and California surveys use such frames). Licenses are linked directly to the angler population of interest, but license frames can suffer from overcoverage (e.g., due to out-of-date licensing information), undercoverage (due to license exemptions or poaching), and duplications. Overcoverage in the license frame is much less than with RDD so sampling is potentially far more efficient. Undercoverage is reduced if license exemptions are minimized. Undetected duplications could be problematic because anglers with more than one license listing may be more avid anglers and would be overrepresented in the sample. Clearly, there is a need for a complete angler registry in all states; these should be designed rigorously to minimize under- and overcoverage. If license frames suffer from substantial incompleteness, then dual-frame approaches could be and should be used to adjust for this incompleteness rigorously, but this will be more expensive and make the surveys more complex than if a complete license (registry) file frame were available.

In the for-hire sector, list frames of operators (based on licenses) are available and being used in telephone surveys in many regions of the country. The same issues of making sure that these lists have minimal under- and overcoverage problems are important.

Frames for Catch per Unit Effort Estimation

The population of interest for CPUE estimation is the population of angler days or trips. This population is, on occasion, accessed through an area frame of coastal waters, with roving boats visiting fishing vessels, but this is expensive and impractical in large spatial-scale surveys and also may be seen as intrusive. As a result, it can be difficult to count and measure the fish caught accurately. The population is accessed primarily through a list frame of site days, where sites are documented fishing access points. (Such site lists are used on occasion for assessing effort, as noted above.) Site days are selected with guidance from a “pressure matrix” that indicates expected fishing intensity across site days. Once a site day is selected, field personnel visit the site on that day and attempt to intercept returning anglers. The field personnel have considerable latitude in how they go about intercepting anglers. Errors in estimating the expected fishing intensity and failure to account for expected fishing intensity in the estimation process can lead to both increased variance and bias in the CPUE estimates.

The major problem with site list frames is undercoverage. Some public access points may be missed in the listing procedure, and private access points are not listed at all. Estimates of CPUE may be biased if anglers accessing the water from private access points or from little-known public access points differ in their fishing (e.g., fishing modes, areas and species targeted, effort and success rate) from those accessing the water from well-documented public access points. In the for-hire sector, access-point interviews of anglers also are required, and the same issue of inaccessible private marinas may apply.

Since expertise on local geography, fishing modes, and species variation is critical, maintenance and sampling of access-point list frames for CPUE estimation is best done at a local level. Even with outstanding local expertise, access-point list frames have a number of potentially serious deficiencies, as outlined above, and need to be supplemented with area samples or other dual-frame techniques to get at CPUE for anglers not accessing the water from listed public access points.

National Registry Frame

This discussion of difficulties with existing frames means that, barring major advances in technology (such as remote sensing) that would allow assessment of fishing effort day by day, a much improved frame for interviewing anglers is needed. Use of the RDD approach in coastal counties is inefficient, potentially biased, and likely to grow even worse over time, but it is the only currently viable option in states without a complete registration of marine anglers to provide a license frame.

A national registry database built on existing state angler licenses and augmented with new licenses would be an ideal frame for sampling marine anglers if it minimized duplications through rigorous and nationally consistent registration standards, minimized overcoverage with regular database updates, and minimized undercoverage by disallowing exemptions. Such a national registry database would yield considerable efficiency for sampling effort over the current RDD frame.

There would be enormous management benefits, cost and interview savings, and increased quality of the catch estimates obtained.

Some states currently require a license to fish in marine waters but do not use the associated angler information to conduct effort or CPUE surveys. This happened in states where the license was developed principally as a means of revenue generation with little application to data collection. Because of the associated fee component, these licenses frequently have numerous exemptions, which reduce their usefulness for frame development and sampling. For example, in Florida,[1] only anglers fishing from a boat in state waters (or traversing state waters to land fish caught in the exclusive economic zone) must buy a Florida saltwater fishing license. Anglers fishing from shore and those over 65 and under 16 are exempt and therefore would not be contacted if the license frame were used for data collection. Saltwater fishing license requirements vary by state, as do the exemptions. Therefore, many current license programs would need to be modified substantially to be suitable as a complete sampling frame.

The recognized need for a national list frame of anglers is not new, and several previous reviews have offered similar recommendations (National Research Council, 2000), but there has been significant resistance from some states to federal involvement in this issue (Box 3.1). Some fear that the additional cost associated with purchasing a license will dissuade people from becoming anglers, and those that are now exempt from license fees likely will resist imposition of the fee if they are required to purchase a license. Further, in the northeast in particular, there appears to be a cultural aversion to the basic idea of saltwater licensing. Still, there are many reasons why a state-level saltwater angler license would benefit data-collection efforts. Cooperation between the federal and state governments on a mandatory salt-water angler registry (or license), with attention to eliminating exemptions in states with current saltwater licensing and with encouragement to other states to implement such licenses as quickly as possible, would lead to realization of those benefits.

The national registry and state survey programs would need additional funding to establish and maintain this type of database. However, there also would be large cost savings associated with sampling from this frame as compared with RDD, where a small proportion of the contacts reach an angler. An updated, complete registration list would greatly

---

[1] Refer to Florida Fish and Wildlife Conservation Commission (2005) for a full list of Florida's exemptions.

Box 3.1
Lessons Learned from Boating Registration

State fisheries agencies generally believe that the federal government lacks sufficient authority for requiring saltwater licenses for those who land their fish in state sovereignty waters. Because of the delicate balance of state and federal interests in marine fisheries, implementing a national saltwater fishing registry continues to be a contentious issue and significant political will may be needed. However, there are important lessons to be learned from recreational boating, most notably, boat registration and numbering. Previous legislative actions for this sector can serve as model for the state–federal cooperation that will be needed in establishing a national angler registration.

At one time, some states had recreational boating registration systems while others did not. Likewise, there were differences in how each state registered boats, the data they collected from owners, and the interval in which information was updated. The Federal Boating Act of 1958 (46 U.S.C. 527-527h) gave states the responsibility for registration and numbering of all undocumented boats after years of benign neglect by the U.S. Coast Guard. Furthermore, national standards for registration and numbering were instituted, including what data were to be collected from boat owners. Deficiencies in boat coverage for numbering and registration purposes were remedied in follow-up federal legislation (Federal Boat Safety Act of 1971 [46 U.S.C. 1451 *et seq.*]). This statute provided incentives in the form of additional funding for states that adopted uniform laws; states that failed to do so were penalized by having a federal numbering and registration system implemented in their respective state.

Thus, to improve the quality and quantity of survey data on marine recreational fisheries, there is a need to establish national standards for existing and proposed state-level saltwater angler licenses or for eventually generating a national universe of marine anglers. It is not automatically necessary to establish a national saltwater fishing license to be administered by the National Marine Fisheries Service (NMFS). There are notable differences here, and the words are important. Some states with saltwater licenses may only have to modify the types of data they collect or expand licensing coverage to anglers previously exempted; other states may need more convincing. Federal standards should deal with the exact types of data collected from anglers and should require that exemptions be eliminated or kept to an absolute minimum.

improve efficiency both in terms of time and cost. It is not assumed that these savings would cover the entire cost of maintaining such a database. However, the benefit from the increased quality and quantity of the data will be well worth the extra cost, especially if there is an associated increase in public confidence with the final estimates. Also, the creation of such a list will be essential to implementing some of the other recommendations found in this report.

It is critical that the licensing requirements eliminate exemptions and noncompliance by segments of the fishing public. Significant efforts to enforce these registration requirements will be necessary. The statistical problems arising from any unavoidable incompleteness of the frame can be addressed in various ways, with the most important one being the use of a dual-frame approach. This will add additional expense so it is crucial to minimize undercoverage of the saltwater license frame. Also, the benefits associated with the angler list frame would be diminished if this list also included freshwater anglers. Including freshwater anglers in the same database would reduce the efficiency gained by the implementation of the registration—unless the data about each angler identifies them either as a freshwater angler, a saltwater angler, or both.

OTHER SURVEY DESIGNS

Panel Surveys

A panel survey is another methodology that has been used in collecting recreational fisheries data. One example is the telephone diary panel survey used in Australia to assess recreational fishing (Henry, 2002; Lyle et al., 2002). This survey used multiple contact telephone interviews to get both fishing effort and harvest rate over a one-year period. Panel surveys should be considered for the telephone survey portion of the MRFSS (National Research Council, 2000). A rotating panel design, with membership in the panel lasting one year (six waves), might be a reasonable approach for the MRFSS.

Panel surveys collect data from the same individuals at regular intervals of time. This design also is referred to as rotation sampling. The main purpose of such a design is that it produces more efficient estimates of change from one time period to the next. To see this, y_t is defined as the parameter of interest at time t (e.g., total fishing effort in a given wave) and y_{t+1} as the total in the next time period. These totals are estimated by $\hat{y}_t$ and $\hat{y}_{t+1}$, and the change ($\hat{\delta}_{t,t+1}$) is:

$$\hat{\delta}_{t,t+1} = \hat{y}_{t+1} - \hat{y}_t$$

The variance of this estimator ($Var(\hat{\delta}_{t,t+1})$) is:

$$Var(\hat{\delta}_{t,t+1}) = Var(\hat{y}_{t+1}) + Var(\hat{y}_t) - 2Cov(\hat{y}_{t+1}, \hat{y}_t)$$

Obviously, the smallest variance will occur when the covariance between $\hat{y}_t$ and $\hat{y}_{t+1}$ is as large as possible. This typically occurs when these estimators are calculated using measurements from exactly the same individuals since one would expect the correlation to be high between an individual's measurements in consecutive time periods.

If there is interest in estimating the combined total over the two time periods efficiently, the opposite sample design strategy would be desired; that is, it would be best if two estimators were used to have the lowest possible covariance. In practice, the best that can be hoped for is to select independent samples each month.

In most real applications, analysts would be interested in estimating both the change and the total for the two time periods, as well as estimates of total for the individual time periods. As a result, it is common to select a design that has partial, but not complete, overlap in sample from one time period to the next. One-level rotation sampling is a design in which a new independent rotation group or panel becomes a part of the sample at each time period, and another (independent) one rotates out of sample. Each rotation group stays in the sample for a number of periods, not always consecutive. The Current Population Survey (U.S. Census Bureau, 2001), for example, employs eight rotation groups, and each group stays in the sample for four months, out for eight, and then in for four again. (Multi-level schemes are an alternative to one-level designs but are not discussed here.) Panel survey design was an active area of research in the 1960s. More recent papers by Wolter (1979), Cantwell (1990), Nieuwenbroek (1991), and Chhikara and Deng (1992) discuss estimation using a rotation design for an area and list frame in a U.S. Department of Agriculture survey.

Besides the advantage of increased efficiency for estimating change, panel surveys provide other benefits. The cost of making an initial contact with and of training a respondent (if that is necessary, as it frequently is in business surveys) is reduced by using the same respondent more than once. There are also disadvantages to panel surveys, including increased respondent burden, which can have a negative effect on response rate. On the other hand, when a respondent refuses to continue after the first interview, there is better information available for imputation than in nonpanel surveys (Lepkowski and Couper, 2002). Another complication in some panel surveys is response bias. For example, it was noted in the U.S. National Crime Victimization Survey that respondents report more crime in their first in-sample period, possibly due to telescoping or remembering memorable events as closer in time than they actually were (U.S. Department of Justice, 2005). As a result, data from the first in-sample period are not used for estimation but are used to help the interviewer determine if any future crime reports are in the reference period or not (Lohr, 1999).

The two major potential advantages of using panel surveys in the MRFSS and other angler surveys are increased efficiency of estimates of change and reduction in cost of acquiring anglers to interview. In terms of the impact these surveys might have on recreational fishing, it seems the former would provide less benefit than some other applications, and the latter could provide more. The increased efficiency of estimates of change comes from a high correlation between measured response in one time period and the next. In a survey like the Current Population Survey, in which the main characteristics of interest are those related to employment, these are fairly stable for most individuals from month to month, resulting in a high correlation. There would surely be some positive correlation between consecutive measurements in the RDD effort estimation, such as between number and types of trips, but it is difficult to guess the strength of those associations a priori.

The most important benefit of using panel surveys would be in time savings for screening to locate fishing households. Cost analyses would be needed to begin with the cost of identifying a fishing household. Experiments on quality of recall (if considering multi-level rotation designs) would be needed and special attention would need to be given to the handling of attrition and the movement of anglers in and out of the telephone frame.

Internet or Web-Based Surveys

Alternatives to telephone surveys need to be considered. Response rates to telephone surveys are dropping due to overuse and suspicion by the general public. There is also the problem of growing cellular telephone use. The committee makes a long-term recommendation that web-based surveys be considered as an addition to telephone surveys. For national and large regional surveys, the committee believes that fishing effort typically will need to be estimated from offsite interviews; therefore, telephone surveys combined with internet and web-based surveys are the only practical option. (However, as noted at the beginning of this chapter, the committee judges that it is crucial to examine carefully the spatial scale of the survey involved, and this will depend on the management unit for the particular fishery. For smaller scales, access–access and aerial–access may be useful design approaches.)

Serious consideration should be given to augmenting telephone surveys with web-based surveys. An internet survey equips a respondent to complete what Dillman (2000) refers to as a computer-assisted personal interview. Internet surveys are based on a random sample of panelists contacted repeatedly (see earlier section on panel surveys); for anglers, this could be from RDD of the complete population or from a license frame. Those without internet access could be provided with the necessary hardware and given free internet access, or they could be contacted by telephone.

Use of the internet would offer a number of advantages, including the ability to handle complex questionnaire skip patterns, to define fishing sites and fish species clearly to respondents, and to deal easily with the reoccurring (panel or diary-type) survey. The labor and time required to contact people continuously by telephone in a reoccurring survey is vastly simplified with an internet survey. One email to your population tells everyone to complete this wave's survey. Another single email reminds them to complete the survey at a later date if they have forgotten. If a person has taken no trips over the past month, he or she simply responds "no trip", and the survey is complete for that wave. If respondents have taken a trip, they are asked, "In which state or states did you go saltwater fishing in the past wave?" They would "click" on a list of all the coastal states shown on the screen. Then the respondents would be asked questions about where they fished and what they caught state-by-state only for states they indicated they had visited. In each "state frame," they would see a map of fishing sites (or counties or coastal areas). They would then indicate which they had visited during the period. Next the respondents would be taken to a "site frame" where they would be asked, "How long did you fish at this site? What did you target at this site? What mode did you use at this site?" In each case, the screen would give choices to "click." For example, mode would show "private boat", "shore", and "charter".

At this stage, respondents also could be asked to report the type and number of fish they caught and released on the trip to validate or compare to onsite information. They would be shown a list of species by name with a drawing or photo of the species. They would "click" on all relevant species caught. Then they would be asked species-by-species (for only the species caught) to report their catch. These surveys also would be a valuable means for getting information on catch-and-release fishing, night fishing, and fishing from private access points that are not covered in the access portion of the current MRFSS. This additional information would allow for some estimate of the biases related to undersampling of those anglers.

Internet surveys also have their difficulties, such as the following:

- Response will be affected by the computer literacy levels of the respondents (e.g., skill with using a mouse and a keyboard, ability to navigate web-type surveys).
- Respondents may be less attentive without an interviewer, which can generate larger response errors.
- Nonresponse rates may be higher or lower (the committee suspects lower due to telephone surveys becoming so unpopular with the public).

Using an internet survey in the for-hire sector to obtain diary information from charter boat owners seems possible and may become routine in the future. Using an internet survey to validate or to compare to kept catch in the onsite access survey also seems very attractive. Further, the committee concludes that internet surveys should be routinely used (perhaps in combination with other modes) in national economic and social surveys (see Chapter 5).

Estimation of Released Catch

The special problems on estimation of released catch (type B2) need to be the subject of more research. As previously discussed in Chapter 2, some of the issues involved are the following: (1) released catch cannot be inspected in onsite survey, unlike the kept catch; (2) rounding errors are common; (3) exaggeration or underreporting due to memory problems are possible; (4) species identification errors may be serious; and (5) the size and age distribution may be different from kept fish. Better methods to estimate the number released are needed. Use of observers in boat-based fisheries to get direct estimates of numbers released could be explored. However, there are problems with releases being different if observers are present. Also, released catch are usually not incorporated into catch estimates, even though there is the potential that high hooking mortality could result in high mortality of the released catch. Released fish mortality estimation from cage studies and tagging studies needs more attention because auxiliary data on depth caught and release condition are hard to collect but are important.

For-Hire Sector Survey Design

For some fisheries, the for-hire sector is responsible for taking most of the recreational catch, which is, in some cases, the majority of the total catch (Coleman et al., 2004). There are at least 10,000 registered charter vessels in the United States. In Alaska, 1,400 charter vessels landed over 60 percent of the reported recreational catch of halibut and lingcod in recent years, with this percentage reaching over 70 percent in southeast Alaska.[2] In the Gulf of Mexico, charter vessels land an average of 70 percent of the recreational red snapper catch (35 percent of the total directed catch), and as a result, a charter vessel moratorium program is being implemented to limit the potential catch from this sector (Gulf of Mexico Fishery Management Council, 2004a). Due to the large potential contribution of this sector to total removals, it is important that it be monitored accurately.

Several years ago, it was recognized that the MRFSS was not effective for assessing the for-hire sector, and consequently, there are now alternative surveys in place in most states for collecting data from the for-hire sector. The most important of these are the For-Hire Survey and the Party Charter Survey (see Appendix B). Both of these surveys are designed to ascertain fishing effort and CPUE data, just as the original MRFSS aims to do. However, the major change is that effort is determined from boat directory telephone lists instead of the RDD frame. Use of these list frames is much more efficient than use of the RDD frame. This allows for a greater sample size specific to this sector. In addition, the potential for bias is eliminated since fishing effort for both

---

[2] Personal communication, Allen Bingham, Alaska Department of Fish and Game, Sportfish Division, Anchorage.

local and nonlocal anglers can be estimated directly from the charter companies. There is no need to adjust for effort by out-of-frame anglers.

The current surveys are capable of monitoring the for-hire sector better than what was achieved through the MRFSS. However, design issues associated with these surveys still exist. The estimation of CPUE still relies on intercept sampling at points of landing; therefore, they are still subject to the problems discussed in the previous chapter about interviewer choice. In fact, intercept issues for this sector may be an even bigger problem since cluster effects arise from multiple anglers participating in the same fishing experience. These effects can be significant and must be accounted for in the estimation for this fishing mode. Another difficulty in surveying the for-hire sector is that operations range from very small to very large, with some being transient. License frames for this sector are likely to suffer from some incomplete coverage, especially for the small or transient operations.

An alternative to the current sampling surveys is the use of mandatory logbooks or diaries of all the fishing effort and catch on for-hire boats, as a condition of the vessel's license. The captain would be responsible for filling out the logbooks as fishing progressed each day, and he or she would be required to turn the logbooks in on a timely basis as a condition for continued licensing. Having the license of the vessel tied to the logbook requirement would be the mechanism to achieve a complete list frame for this sector. Therefore, a census of this sector theoretically is possible because the population of charter and head boats is defined more easily than that of the total angler population. Also, the captains and crew generally have a greater knowledge of the local fish species and could provide more reliable catch data, including species identification and the location of catch.

The question of whether to use a survey or census for the for-hire sector is not a new one. In 2001, the Recreational Technical Committee (RTC) of the Atlantic Coast Cooperative Statistics Program (ACCSP) undertook a one-year assessment of three programs designed to measure the fishing activity of the for-hire sector of the South Carolina marine fishery (Ditton et al., 2002). The purpose was to provide information for determining the best and most acceptable method of collecting data from the for-hire sector that could be adopted as a standard by ACCSP. They reviewed (1) the MRFSS; (2) the mandatory South Carolina Charter Logbook Survey, combined with the NMFS Headboat Logbook Survey; and (3) the NMFS Vessel Directory Telephone Survey (VDTS), combined with the MRFSS intercept component (with augmented sampling)—the precursor design for the For-Hire Survey.

RTC noted marked improvements with VDTS and the South Carolina logbook methods when compared to the MRFSS (Ditton et al., 2002). RTC found that the advantages of the logbook program were that it had the most credibility of the three methods with the public, it had the best timeliness of data availability, and it had the most complete sampling frame and coverage. It also found that the logbook program sampled 99 percent of the for-hire vessels in South Carolina and was successful because it was mandatory, enforceable (with measurable enforcement actions), and financially sustainable. The disadvantages of the South Carolina Charter Logbook Survey were the possibility of an incomplete sampling frame because of potential rogue vessels, underreporting on vessels, and lack of biological sampling dockside. RTC provisionally recommended the VDTS program over the logbook program because of implementation issues it anticipated for a coastwide program—primarily lack of funding and commitment of agencies to enforcement and validation. At least some of the potential problems identified for the South Carolina Charter Logbook Survey could be addressed through a logbook–license linkage, as described above.

Although RTC was concerned that there could be implementation issues for logbook programs, this will be true with any fundamental change in sampling protocol. Given the magnitude of the for-hire sector in some regions and the potential scale of fishery removals for this sector, the committee finds compelling arguments for the use of mandatory logbooks as the source of catch and effort data for the for-hire sector. Furthermore, ACCSP found that recreational and charter fishing constituents along the Atlantic coast have a strong desire to participate more actively in data collection (Loftus et al., 1999), and the committee heard similar comments during public testimony.[3] Not only can mandatory reporting from the for-hire sector increase the public's acceptance of the credibility of recreational catch statistics, it may help to facilitate "ownership" of these data by the for-hire sector. If the data they are supplying are a component of the final estimation, there may be fewer criticisms of these final estimates.

The use of logbook data is particularly important for fisheries in which fishery-independent surveys are conducted infrequently or not at all because these data will be an essential component of stock assessment calculations. The committee recognizes that logbooks should not be re-

[3] The testimony from a small number of individuals on this topic may not represent the whole fishing community; however, this testimony originated from the largest national organization of charter boat operators.

quired by more than one level of government (state, regional, federal, and international), and agencies must be coordinated to avoid the burden of duplicate reporting. The committee sees the state as the appropriate level of implementation for this requirement, with adherence to national reporting standards and program coordination at the national level.

Validation of data acquired through any source is an issue of concern, and it would be no less so for a mandatory logbook program. The data collected through logbook programs will be reliable only if there are strict verification and enforcement components of the program. Since the information obtained from the logbooks is owner supplied, there is the need for verification for both CPUE and effort. Effort and kept catch could be checked by dockside inspection of angler parties and their catch. However, accurate and timely logbook submission as a condition of license is important. While the normal process of validation through creel surveys and random sampling of individual clients on the vessels could still be used, there would be direct and effective accountability because of the legal requirement for the logbooks, as well as the economic incentive associated with continued licensing of the charter operation. Also, the logbook program will serve as a participation record for any more detailed allocation discussions (e.g., the use of individual quotas for charter vessels, which is being contemplated in some jurisdictions). Finally, a mandatory logbook program provides a comparison vehicle for data acquired independently via offsite, random, individual angler-based or panel-based surveys.

A for-hire logbook program represents a significant step in monitoring of this sector, but it will not solve all problems of monitoring. For example, accurate accounting and verification of catch-and-release activity will be addressed only partially through such a program. Alternative verification of catch and release via observers or electronic monitoring may be required. However, the committee views the for-hire sector as a business enterprise—the business being the connection of people and fishing opportunity. Therefore, this sector should be subject to a greater level of reporting than independent anglers, as a corollary of conducting business based on a public resource.

Such a program will require additional resources to maintain a logbook-based data infrastructure. However, the substantial benefits of the program, as recognized in previous reviews and some existing programs, argue for its adoption and the commitment of resources to its implementation. Also, there are significant design issues associated with stratification by size of charter operation and geographic locality. These design issues are further addressed in Chapter 6.

ANALYSIS AND ESTIMATION TECHNIQUES

Below is a detailed overview of analysis and estimation issues related to the MRFSS that could revolutionize the way the survey is analyzed, especially at smaller spatial scales. It is deliberately presented at a higher technical level than some of the other sections because of the complexities involved. Generally, the analysis issues are focused on the use of auxiliary information to increase precision and the special problems with estimation of subpopulations. In virtually all surveys, estimates are required not only for the population as a whole but for various subpopulations, called *domains*. For human populations, domains may be demographic groups (e.g., age, race, sex), occupational groups, or geographic groups. For natural resource inventories, domains are typically geographic (e.g., county, state, state waters, federal waters) or ecological subdivisions (e.g., ecoregion, watershed). Geographic subpopulations are called *areas*. In fisheries, domains could be geographic areas or temporal periods.

Often domains are not sampling strata so the sample size within domains is not pre-allocated but is determined randomly from the sampling. Three useful classes of domains are *large domains*, *medium domains*, and *small domains*, based on the sample sizes attained in those domains.

Large Domains and Direct Estimation

Large domains are likely to be sampling strata (i.e., predefined subpopulations that are sampled independently using predetermined sample allocations), but even if they are not, they are large enough to have a high probability of a large sample size. This large sample size ensures that standard design-based survey estimation procedures yield estimators of adequate precision. These standard estimators are called *direct estimators* because they use data only from the study units in the domain and time period of interest. These estimators have good design properties, and they are typically unbiased (or asymptotically unbiased), asymptotically normal, and allow for statistically consistent variance estimation and valid confidence intervals. All of these good statistical properties are justified by the randomization used in the probability sampling design and do not depend on the validity of any statistical model. This is the approach usually used in the current MRFSS analyses.

Medium Domains and Survey Regression Estimation

Direct estimation is not reliable if the sample size is too small. In medium domains, the sample size is moderate but not extremely small. For such domains, if auxiliary information is available at both the population level and the sample level, it is often possible to construct a *survey regression estimator* (e.g., Cochran, 1977) with greater precision than that of the simple direct estimator. Such an estimator fits a global regression model to all of the survey data and predicts the responses for unsampled population elements using the fitted model. Survey regression estimators may be either *model-based* or *model-assisted.* Model-based survey regression estimators estimate the total for a domain by adding the responses for the sampled elements to the predicted responses for the unsampled elements. Such estimators are highly efficient if the model is right but can be biased and even inconsistent if the model is wrong. On the other hand, a model-assisted survey regression estimator predicts all elements using the fitted model and adds them up over the domain of interest. Since this prediction may be biased if the model is not specified correctly, the model-assisted estimator adds on a design-bias adjustment computed as the weighted difference between the observed and predicted responses over the domain. If the model is right, the estimator is highly efficient. The key result is that whether or not the model is right, the model-assisted estimator retains the good design properties of a direct estimator (i.e., it is asymptotically unbiased, asymptotically normal, and allows for consistent variance estimation and valid confidence intervals) (Särndal et al., 1992).

The type of survey regression estimator depends on the types of available auxiliary information. With categorical covariates only, the survey regression estimator is a *post-stratified estimator.* With a single continuous covariate, the survey regression estimator could be a *ratio estimator*, *classical regression estimator*, or even a *kernel* or *spline-based nonparametric survey regression estimator* (e.g., Breidt and Opsomer, 2000). Generalizations to multiple covariates are also possible.

To ensure the quality and timeliness of any of these survey regression estimators, all covariates that enter the regression must be of high quality and must be readily available in a timely manner. Definitions of the covariates and protocols for their measurement should change as little as possible over time. Missing covariate information should be minimal. Indeed, all of the quality standards applicable to responses in the original survey are applicable to the covariates as well.

In the context of fisheries surveys, possible covariates for effort could include business-related covariates (e.g., bait sales, boat rentals) and weather-related covariates (e.g., precipitation, temperature, wave height). The business-related covariates could be difficult to obtain and use in an ongoing survey. Establishments vary in size and in the resources they devote to maintaining accounting records. Quality can vary considerably from establishment to establishment and from year to year. Definitions would need to be standardized, and cycles of data compilations would need to be synchronized. Thus, tracking down and compiling sales or rental data could be as difficult as conducting the original survey. The weather-related covariates, on the other hand, are readily available in a timely and consistent manner from a centralized source—the National Oceanic and Atmospheric Administration itself. Use of these weather data should involve minimal additional cost.

Small Domains and Small Area Estimation

The final domain classification is the small domain, called a *small area* in a geographic context (Ghosh and Rao, 1994; Rao, 2003). Here, direct estimators or model-assisted survey regression estimators are not sufficiently precise for the inferential problems of interest. Typically, the random sample size in a small domain or area is small and may be zero in some cases. There is no hope for direct estimation with such small sample sizes so small domain estimation problems lead to *indirect estimators*. Unlike direct estimators, indirect estimators use data from outside the domain or time period of interest to "borrow strength" across time or space, and the validity of these indirect methods depends on the correctness of the model specification.

Perhaps the simplest small area estimator is the *synthetic estimator* in which all elements in a domain are predicted from a fitted global model relating the response variable to the covariates. The model borrows strength from the entire sample in the fitting of the regression model, which typically has common coefficients for all domains in the population. The synthetic estimator can be computed for a given small domain even if there are no samples in that domain and usually has very low variance since it is fitted on the basis of the entire sample, but it may have large bias if the model is incorrectly specified.

A *composite estimator* attempts to trade off the low bias but high variability of a direct estimator with the high bias but low variability of a synthetic estimator by computing a combination of the two estimators.

The weights in this composite estimator can be chosen in an ad hoc way, such as by making the weight on the direct estimator larger if the sample size in that domain gets larger. The weights also can be chosen on the basis of a formal statistical model. The standard approach to formal composite estimation is to choose the composite weights as functions of the parameters from a fitted model. Two classes of models appear in the literature, depending on the type of available auxiliary information. *Element-level models* require auxiliary information for every sampled element (e.g., Battese et al., 1988), while *area-level models* require auxiliary information only for each small area. In either case, the small area model is hierarchical. In area-level models, much of the complexity of the survey design is averaged out, and nonnormality in responses tends to average out as well.

Here, the focus is on the area-level model. Assuming that auxiliary information is available for each small area, the model describes the distribution of the direct estimates given the true domain parameters, and the distribution of the true domain parameters given the covariates. Usually, the direct estimate is modeled as truth + sampling error where the sampling error has a mean of zero and known variance. The true domain parameters are modeled with a global regression function of the covariates, plus domain-specific deviations from the global model. The domain-specific deviations are random effects that may have some correlation structure, such as temporal correlation structure in a time-indirect context or spatial correlation structure in a domain-indirect geographic context.

The small domain model has two ways to borrow strength: globally through the regression fitted to all the data and locally through the temporally or spatially correlated random effects. Temporal correlation structure can be described with a *state space model,* special cases of which can include *autoregressive moving average models* (e.g., Brockwell and Davis, 1991). State-level unemployment estimates from the Current Population Survey, for example, combine a regression model, a basic structural model for stochastic trend and seasonality, and an autoregressive moving average model for the correlated sampling errors (Tiller, 1992). Spatial correlation structure in an area-level model can be described with a *lattice model*, such as a *conditional autoregression model* (Cressie, 1993).

Small area estimation models are fitted using standard statistical procedures, such as through estimation of variance–covariance parameters by restricted maximum likelihood or other methods followed by joint estimation and prediction of the fixed and random effects in the model. This approach is known as *empirical best linear unbiased prediction* and has relatively straightforward computation that may be implemented using standard statistical software (e.g., the PROC MIXED function in SAS software, the lme function in S-Plus software), but these methods do not account fully for uncertainty since they treat the estimated variance–covariance parameters as known. *Hierarchical Bayesian analysis* is also possible. Prior distributions for all unknown parameters (including variance–covariance parameters) are assigned, and then numerical techniques, such as Markov Chain Monte Carlo, are used to compute posterior distributions of the unknown parameters given the direct estimates. Computation is more complex, but now these methods are routinely taught to statisticians in graduate school and routinely implemented in many government agencies that employ statisticians.

Use of These Techniques in Angler Surveys

In the context of angler surveys, use of auxiliary variables and small area estimation techniques might be applied to the effort estimates, the CPUE estimates, or the final catch estimates, perhaps after some transformation. Suitable auxiliary information for effort modeling may include weather-related covariates; suitable auxiliary information for CPUE may or may not be available. Identification of suitable covariates and specification of an appropriate regression model or models would be a critical part of a small area analysis.

Even without suitable covariates, estimation of both effort and CPUE might be assisted by temporal, spatial, and multivariate correlation. The data are collected in temporal waves, and wave-to-wave or year-to-year correlation might be helpful in predicting current wave values. Also, the data are spatially explicit, so borrowing information from similar, nearby areas might help to improve predictions. Finally, the data are multivariate (catch by species), and the correlation structure among the different species might help in predicting individual species components.

To conclude, the current estimation methodology used in the MRFSS is primarily direct estimation for large domains. Auxiliary information and survey regression estimation methods enter in minor ways, such as in some simple ratio adjustments and temporal pooling of estimators. It appears that with relatively modest additional resources, the MRFSS could add more formal survey regression methods, extending the inferential scale to medium-sized domains. Small area estimation would require a much greater investment of resources. This estimation methodology would require stronger assumptions, more sophisticated model specification (both in the regression model and in the covariance structure), more detailed diagnostics, and heavier computations. However, the potential pay-off is enormous in that it extends the inferential scale to finer spatial resolutions, which seems to be what managers currently require. These recommendations will require a rethinking of the program management of angler surveys (see Chapter 6 on program management and support).

CONCLUSIONS AND RECOMMENDATIONS

The committee concludes that the current methods used in the MRFSS for sampling the universe of anglers and for determining their catch and effort are inadequate. Sampling of each group of anglers (i.e., private, guided, head boat, and charter boat) presents challenges that can differ across the groups. Two complementary methods of sampling angler catch and effort are used in the MRFSS. One is onsite (i.e., intercepting anglers while they are fishing or at their access [landing] points). The other is offsite, which includes a variety of sampling techniques for contacting anglers after they have completed their trips. Both onsite and offsite methods suffer from weaknesses that may lead to biases in catch and effort estimation. This necessitates major changes in both the design and analysis procedures.

A comprehensive, universal sampling frame with national coverage should be established. The most effective ways to achieve this are through a national registry of all saltwater anglers or through new or existing state saltwater license programs that would allow no exemptions and that would provide appropriate contact and information from anglers fishing in all marine waters, both state and federal. Any gaps in such a program (e.g., a lack of registration in a particular region or mode, exemptions of various classes of anglers) would compromise the use of the sampling frame and, hence, the quality of the survey program. **Future telephone surveys should be based on the above universal sampling frame.**

Dual-frame procedures should be used wherever possible to reduce sample bias. For example, if a state has an incomplete list frame based on licenses, the use of an additional sampling frame of the state's residents (e.g., RDD) would reduce the bias. The existence of a universal frame described above would make this approach unnecessary for offsite sampling, provided there are no exemptions. **Complemented surveys**

should be used more widely in regional surveys where reliable estimates are required for management of a small suite of very important species at small regional scales.

Panel surveys, which contact individual anglers repeatedly over time, should be considered in recreational fishing surveys to gather angler trend data and to improve the efficiency of data collection. This is especially true for the telephone portion of the MRFSS.

Internet surveys should be considered for their potential use in recreational fishing surveys, especially in panel surveys, as a way for anglers to submit information. They could be used in the for-hire surveys, in private angler surveys like the MRFSS, or in social and economic surveys.

In most cases, **charter boat, head boat, and other for-hire recreational fishing operations should be required to maintain logbooks of fish landed and kept, as well as fish caught and released.** Providing the information should be mandatory for continued operation in this sector, and all the information should be verifiable and made available to the survey program in a timely manner. Onboard observers could be used on a sample of vessels to verify logbook information. A sample survey may be more appropriate in fisheries where the for-hire sector is a small component of the catch or where verification and enforcement are particularly problematic.

The reported release alive of captured fish (catch and release) is increasingly common in many marine recreational fisheries. Although released fish suffer lower mortality than retained fish (the mortality of retained fish is, of course, 100 percent), there still is some mortality, and in some cases, it can exceed 50 percent. **The survey fails to provide a valid and reliable method of adequately accounting for fish caught and *not* brought to the dock (including fish released alive or dead, as well as fish caught for bait or given away before reaching the dock). This shortcoming affects estimates of catch and total removals.**

Current analysis procedures used in estimation for the MRFSS do not exploit the current knowledge of finite population sampling theory. The current estimates are particularly deficient when applied to small areas because they do not use information in adjoining areas or time periods, nor do they consider relationships between species that occur together. Therefore, they are of lower precision than would be possible if this information were used. Improvements in these estimates would be of great use to managers who need to make quick decisions concerning spatial areas that are smaller than typical in the early years of the MRFSS.

4

Data Requirements for Population Assessment

Three kinds of data are required in developing population assessment models: (1) total catch (or total removals if including bycatch); (2) demographic information on the size, age, and taxonomic composition of the fish removed; and (3) indices of relative abundance. Total removals by size and age are used to measure the level of mortality incurred by different components of the population. Abundance indices serve to denote relative change in the fish population over time. These indices can be based on data collected directly from the fishery (i.e., fishery-dependent indices, such as catch rate indices from fishery logbooks) or data collected independent of the fishery (i.e., fishery-independent indices, such as research surveys). Information on these three kinds of data ideally should be obtained from all fisheries and gear types involved in removals from the population.

Commercial fisheries have been the main source of fishery-dependent data used in developing quantitative population assessments; however, more and more often, data from recreational fisheries are relied on to complement data collected from other sources or as the sole source of information for some assessments. This increased demand on recreational fisheries data necessitates a discussion of the survey methods used in recreational fisheries and whether these methods provide data adequate for assessment purposes.

Certainly the survey design and data collection recommendations outlined in earlier chapters are likely to improve the information used for population assessments. In particular, the establishment of mandatory logbooks to monitor catch for all vessels in the for-hire sector would provide more in-depth data—the kind of data that would be ideal for use in population assessments. Logbooks could provide fishing location, time of day, and weather conditions, all of which could be helpful in interpreting catch rate estimates. Onboard observers also could be used for one-time studies into catch rate, such as investigations into the influence of the kind of bait used or the depth being fished, or they could assist in the collection of specific oceanographic and meteorological data during the fishing trip.

In addition to identifying reliable data sources, data quality must be assessed and accounted for appropriately. Modern statistical population assessment models are capable of dealing with data characterized by different variance structures, or even unknown variance. Not surprisingly, what goes into the model influences what comes out, and the accuracy of population estimates is influenced by the accuracy of the data used. Statistical fitting procedures used in these models often assume variance structure for data inputs that are not likely to be met by most recreational fisheries sampling programs. Assessment models can be modified to accommodate such data characteristics, but these characteristics first must be identified and quantified at the source level of the surveys.

Inconsistencies in how dockside samples are collected can be particularly aggravating when conducting population assessments. For example, the lack of a common knowledge base among anglers, data collectors, and data users with regard to taxonomic identification will bias mortality estimates for all species concerned. Population assessment scientists must have confidence that species designations are accurate and applied consistently in the sampling process. Therefore, biological data obtained from intercept surveys must be consistent with categories used in assessments.

Two additional issues complicate the usefulness of recreational fisheries data for population assessment. One concern is the challenge faced by population scientists in interpreting catch and effort data recorded from recreational fisheries surveys in ways that are analogous to commercial and scientific indices to measure changes in relative abundance. Obtaining a measure of catch per unit effort (CPUE) that is a true measure of relative abundance is challenging since the measures for these different data sources are compiled with different purposes in mind. For commercial fisheries, catch and effort are obtained simultaneously from individuals in association with an area fished and species targeted; thus, CPUE can be seen as a direct measure of relative abundance for a given area and species, as long as fishing efficiency and catchability do not change. In recreational fisheries surveys, such as the Marine Recreational Fisheries Statistics Survey (MRFSS), CPUE is obtained from individuals and is expanded by an estimate of effort across all individuals to develop an estimate of total catch. For these surveys, recreational CPUE typically is not associated with specific areas or even with specific target species; thus its applicability as a relative abundance measure is clouded by its design as a means to obtain total catch in the survey. To better address these issues, a closer look needs to be taken at effort and CPUE calculations as they are carried out in a recreational fisheries context.

The other complicating issue is how catch and release influences the accuracy of total removals reported and the subsequent underestimation of fishing mortality. Underreported removals occur when fish are released but subsequently die from capture and handling. If catch-and-release survivorship rates are not known, the proportion of releases that die is not known. This is further complicated by the fact that the number of releases (by species) probably is not estimated accurately either. The numbers of released fish are obtained from the intercept survey, and the accuracy of this information may be dependent upon the memorableness of the release event. Most anglers would remember releasing a marlin but may be uncertain as to how many of a more common species, such as striped bass or mackerel, were released.

EFFORT AND CATCH PER UNIT EFFORT CALCULATIONS

Stock assessment scientists often use the reported CPUE from commercial logbooks as a fishery-dependent index of abundance. The basic assumption is that catch (C) is a function of fishing effort (E), and catchability of the fish to the fishing gear used (q) is constant over time, such that:

$$C = f(E,q,N)$$

where N is the population size. Effort is usually a function of time spent actively fishing (trawling) or the time a specific amount of passive gear (e.g., traps, pots, longlines) was in the water. The simplest form for this function is a proportional relationship:

$$C = E \times q \times N$$

and as a result:

$$CPUE = \frac{C}{E} = q \times N$$

Assuming a constant catchability, CPUE should track changes in population size over time.

In the MRFSS, the catch per trip from the onsite interview survey is often referred to as a catch rate or CPUE, but this estimate is rarely the one used in stock assessments when defining a recreational CPUE index of abundance. The definition of fishing effort, and hence, fishery catch rate as an index of abundance needs to take into account what species the effort was directed for and not just the total catch over a set amount of time.

Holiman (1996) defines three types of effort that can be calculated from the MRFSS data: target effort, catch effort, and directed effort. Population assessment scientists must be aware of the presence of these three different types of effort in the database and understand how they relate to the problem of estimating relative abundance. Those in charge of data collection and monitoring also must be aware of these effort types in order to document them properly but also to insure that the right type of information can be made available to those in need of it.

Target effort is based on the anglers' identification of their primary or secondary target species to the intercept interviewer, whether or not they were successful in catching any of that species. Interviews occur after the fishing trip is completed; therefore, accepting the angler's designation of target species after the fact may result in biased estimates, as some people may report only what they caught as being what they targeted. This is often referred to as "prestige" bias since it is a result of anglers not wanting to admit that they were unsuccessful in catching what they were targeting.

In cases where there are multiple anglers (e.g., head boats) and catch cannot be separated by angler, total catch is attributed to one angler (termed "leader" in the MRFSS) who represents the other anglers on the trip (designated as "followers"). Generally, it is assumed that all followers fish when the leader fishes for the target species (Holiman, 1996). If followers do not fish when the leader does, the amount of target effort will be overestimated. However, if the leader does not report a target species but one of the followers does, it is not assumed that all followers also targeted that species. This assumption may result in an underestimate of target effort. The most recent Atlantic bluefish stock assessment used target effort to define catch rate indices (Lee, 2003) and therefore might suffer from the difficulties mentioned above. In addition, for schooling fish species (such as bluefish), there is an increased probability that if a follower reports targeting that species, the other anglers did as well.

Catch effort is the effort associated with the successful catch of a species, whether or not it was targeted. In addition to the issues raised above for target effort, assumptions have to be made when calculating the total effort for groups of anglers. Again, all catch, either kept for interviewer inspection or not available (i.e., filleted, released dead or alive, given away), is attributed to the leader. However, the number of angler trips associated with catching these fish is not recorded. That is, assuming that bag limits exceed one fish per angler, one person may have caught more than one of these fish. For example, Holiman's code assumes that if the number of fish is less than the number of anglers, then the number of trips equals the number of fish because the focus here is the successful catch of a specific species; otherwise, the number of trips equals the number of anglers. Effort calculations for the red grouper assessment simply use all of the anglers when dealing with multiple angler intercepts (Southeast Fisheries Science Center, 2002).

Directed effort is the effort associated with all catch of a particular species whether targeted (including unsuccessful catch) or not. The difference between target effort and catch effort is referred to as effort associated with incidental catch.

All of the above deal with effort estimates obtained from anglers interviewed during intercept surveys. However, there is no information available in the MRFSS on target or other kinds of effort for anglers who have private access. At present, it must be assumed that this portion of the angler effort is represented adequately by the sampled portion from intercepts.

Most stock assessments try to use some form of target effort, and the main issue is how to calculate the "target but no catch" portion of the effort in a way that does not rely on the anglers' identification of target species. Ralston and Dick (2003) use location data from the California commercial passenger fishing vessel (CPFV) data to restrict black rockfish data to only those locations where black rockfish had been caught in at least five separate locations. The latest assessment for red snapper (Gulf of Mexico Fishery Management Council, 2005a) uses only trips where at least one snapper was caught or where a catch of species typically associated with snapper was caught in the past. No-catch reef trips for hogfish are defined as reef trips using hook and line or spear in counties where hogfish were not caught in the current trip but had been caught at least once in the period 1982–2001 (Ault et al., 2003). As another approach, Stephens and MacCall (2004) describe a logistic regression approach based on multispecies presence–absence information that predicts the probability that the target species would be present based on other species caught, even if no catch of the target species was recorded. Methods that depend upon species-complex indices to determine targeted non-catch trips may be confounded by differential targeting of the fishery from year-to-year (or season-to-season) on more desirable species or by changes in the complex arising from different dynamics of the component species. Also, regulation changes for associated species may complicate interpretation of species-complex catch information. While all of these methods are genuine attempts to measure target effort, confidence in their use in stock assessments will require more research with respect to multi-species associations and the impact of species-specific catch regulations.

In recent west coast Stock Assessment Review Panel (STAR Panel) reports (Pacific Fishery Management Council, 2006), the use of recreational catch rate data from CPFV logs or Recreational Fisheries Information Network (RecFIN) data for rockfish and similar species has come under close scrutiny. The usual assumption of proportionality between catch rate and abundance used for commercial indices has not been tested for recreational fisheries. These reports note that recreational fishing, especially when conducted by the for-hire sector, focuses on giving the angler a successful fishing experience with respect to the desirability of and the challenge of landing the species being sought. This behavior may lead to the targeting of fish in high density areas, resulting in catch rate indices exhibiting a slower decline than what the actual population is experiencing (i.e., hyperstability). There are many other factors also at play in determining what makes a successful fishing experience (Holland and Ditton, 1992) that may further complicate the link between recreational catch rate and population size. Technological improvements (e.g., Global Positioning System [GPS]) are not usually taken into account when using recreational catch rates as indices of abundance. In addition, changes in fisheries regulations for the targeted or associated species may change the relationship between catch rate and population size. As fisheries become more restrictive with respect to bag and size limits, the increasing number of releases may result in CPUE being prone to recapture bias (e.g., lingcod) (King and Haggarty, 2004).

For many fisheries where there, currently, is a small or no commercial component (e.g., rockfish on the U.S. west coast), recreational catch rates are usually the only abundance indices available for the recent years. Despite the problems that have been identified with using recreational effort data, assessment scientists need to have access to the best effort data possible either to determine whether these kinds of data can be used to monitor abundance or to make the necessary modifications so the data are useful.

CATCH AND RELEASE

For stock assessment purposes, the total number of fish removed from the population by the fishery is of more interest than just the number of fish landed. Total removals are calculated as the sum of those fish caught and landed or known to be dead upon capture and those fish that were released (or discarded) but did not survive. There are two types of catch records in the MRFSS database: type A and B. Type A records account for fish that were caught, landed whole, and available for identification by the intercept interviewers. These fish are available for weight and length measurements, although these measurements may not always be taken. For type B records, the fish were caught but were either not kept or were unavailable for identification. These records are further identified as either type B1 or B2. The former type refers to fish that were filleted, released dead, given away, or disposed of in some way other than for types A or B2. Those fish that were caught and released alive are coded as B2. Total landings from the recreational fishery are calculated as A+B1 for stock assessments where there are recreational components. For example, in 2003, recreational landings (A+B1) in the striped bass fishery were estimated at 2.4 million fish or 11,486 metric tons (25.3 million pounds) from the MRFSS. These landings constituted 74 percent of the total landings of striped bass by the recreational and commercial fishery (Atlantic States Marine Fisheries Commission, 2004).

Some fish released alive, as recorded in the B2 records, are expected to die after being released. This subsequent mortality is often referred to as a hooking or release mortality and can arise for a number of reasons, including swim bladders expanding too quickly as a result of fish being brought up from significant depths. There is also the possibility that fish exhausted from fighting the angler are more susceptible to predation. In 2003, for striped bass, the B2 catch was estimated at 14.6 million fish. Assuming an 8 percent hooking mortality rate, catch and release resulted in an estimated removal of 1.2 million additional fish.

The issue of hooking or release mortalities has been the subject of a number of studies (e.g., Lucy and Studholme, 2002; Policansky, 2002), but these mortalities have not been estimated for all recreational species. Specific studies have often limited their attention to particular combinations of factors, such as hook type, depth range, and temperatures, that can be manipulated in an experimental setting. Not all factor combinations that could be significant may have been studied, but more to the point, there may not be enough detail on the released fish from the intercept surveys to determine which, if any, of these factors are operating at any one time. Although some hooking mortality studies appear to be reliable for the restricted conditions they apply to, not all are reliable, and for many species, the hooking mortality associated with particular gear types under a variety of conditions simply is not well known. In addition to providing information for stock assessment scientists, hooking mortality studies also have the potential to improve management by advising anglers on how to handle and release hooked fish to increase their chances of survival.

The lack of accurate information for estimating release (or discard) mortality has been identified as problematic for the red snapper assessment in the Gulf of Mexico. For red snapper, recreational data were obtained from three sources:

- The MRFSS (1981–1998) with some exceptions: (1) no wave 1 data in 1981, (2) no Texas boat mode in 1982–1984, (3) no Texas data after 1986, and (4) no head boat sampling after 1985
- The National Marine Fisheries Service's Beaufort Laboratory head boat survey for all states after 1985
- The Texas Parks and Wildlife Department's coastal sport fishing survey

Recreational discards data are collected by the MRFSS in the Gulf of Mexico but are not available for Texas landings or for landings from head boats. Mortality rates used for discarded live red snapper differ according to depth and area and therefore depend upon accurate location information of where the discards occur across the whole range of the fishery (see Appendix C).

Many stock assessments convert numbers caught to weight caught. Weight conversions are based on the length and weight information obtained from the type A catch; the size compositions of the type B1 and B2 catch are assumed to be similar to the type A catch—a potentially strongly biased approach given that one of the main reasons for releasing fish is that they are below the size limit. At present, there are some limited programs to capture length, weight, or age data from the recreational discards (e.g., striped bass lengths are available from volunteer angler logbooks and American Littoral Society data), and, starting in 2003, California Recreational Fisheries Survey (CRFS) samplers have measured length and weight of discarded fish from CPFVs and from onshore anglers. (See Appendix B for more detail on CRFS.)

Another problem is the comparability of discard data between different recreational surveys that may be combined into a stock assessment. As an example, recreational catch data for lingcod on the Pacific coast come from a variety of sources, but not all sources provide the same level of detail with respect to the condition of the fish caught or released. For California, the RecFIN database (including the MRFSS) was used for 1980–1989 and 1993–2003. Beginning in 2004, CRFS has been used in place of the MRFSS in California. Oregon recreational catch data are provided by the Oregon Department of Fish and Wildlife, and Washington catch data are obtained from the Washington Department of Fish and Wildlife (WDFW) Ocean Sampling Program (see Appendix B). Discard information on numbers and disposition (released alive or dead) is available from CRFS. On the other hand, only the number released is available from the Oregon Recreational Boat Survey data (see Appendix B). The WDFW has collected discard information from the recreational fishery since 2002 but does not collect information on the portion of the catch discarded live or dead. In Washington, 57 percent of the lingcod catch is estimated to be discarded, but it is unknown how many of the live releases survive. Various adjustments are made to the catch and projections in the assessment to account for discard mortality. Yet, recent stock assessments for lingcod identified the need for better coastwide enumeration of at-sea discards and mortality of released recreational fish to account for total removals from the population more accurately.

CONCLUSIONS AND RECOMMENDATIONS

Documentation of the source of the effort available in the MRFSS, as to whether it falls into the category of target effort, catch effort, or directed effort, would go a long way in helping population scientists use this data in an appropriate manner. Population scientists should work in collaboration with those involved with data collection and monitoring to design data collection protocols and summary statistics that are appropriate for use in population assessment. Information on target species and area fished also would make catch and effort data collected from recreational fisheries more amenable for use in the development of assessment indicators.

The establishment of mandatory logbooks to monitor catch for all vessels in the for-hire sector also would be appropriate for the collection of target effort data. Basic data recorded by the vessel captain on the number of anglers, actual hours spent fishing, and target species would get around the complications of the leader–follower designations currently being used in the MRFSS. The logbooks also could record position, time of day, and weather conditions, all of which could be helpful in interpreting catch rate estimates.

These logbooks would not be considered the sole source of information, and similar to the commercial fishery, **onboard observers should be used on a sample of the vessels to validate the information, especially in the case of numbers, species, condition, and size composition of the released fish.** Recall bias of released fish has been identified as an issue in recreational fisheries (Pollock et al., 1994) and shown to be significant for salmon fisheries in the Strait of Georgia (Diewert et al., 2005). These observers also could be used for one-time studies into catch rate, such as investigations into the influence of the kind of bait used, depth being fished, and discard mortality, or perhaps they could assist in the collection of specific oceanographic and meteorological data during the fishing trip.

Information on targeted effort, such as discussed for the for-hire sector, could be obtained for private access anglers as part of a panel survey. Panel surveys could be used to collect a wide range of detailed data from the previously unsampled private access mode. Participants could be contacted by telephone or as part of an internet survey. It may be possible to design these panel surveys in a way that detailed information on catch rate and targeted species can be related back to the larger telephone survey of private sector anglers providing fisherywide or regional estimates of catch rate for stock assessments. These surveys also can be used to collect information on the sizes of kept or released fish. This may require significant training to ensure accurate species data, but since data will be collected from each participant over a long period time, this investment in training may be worthwhile.

5

Human Dimensions

Especially in recent decades, fisheries management considers ecological, political, economic, and sociocultural factors. From the national standards of the current Magnuson-Stevens Fishery Conservation and Management Act (16 U.S.C. 1801 *et seq.*), it should be readily apparent that fisheries management decision making, whether for recreational or commercial fisheries, requires a diversity of valid and reliable data well beyond "estimating the impact of recreational fishing on marine resources" (the stated purpose of the Marine Recreational Fisheries Statistics Survey [MRFSS] and other National Marine Fisheries Service [NMFS] surveys) (National Oceanic and Atmospheric Administration, 2005b). Most available data are biological and ecological in orientation. The political domain has been dominated by established rules and regulations, by policies of agencies and administrations, and by the values of those involved in the task of fisheries management. There has been a paucity of data on human dimensions available to decision makers in fisheries management.

Part of the lack of data on human dimensions flows from a lack of recognition among fishery managers of the importance of those data, and part has to do with agency tasks. Management councils are tasked with conservation first (i.e., identifying available yield) and optimum use second (i.e., how to best serve the nation with the yield available). This has dictated the priorities of recreational surveys, but the surveys have not evolved along with advances in sociocultural and economic information. The difficulties facing fishery management agencies are as often sociocultural and economic as they are biological, and failure to incorporate sociocultural and economic information into fishery management increases the likelihood of management failure. In the statement of task, the committee was asked to consider the match or mismatch between options for collecting recreational fisheries data and alternative approaches for managing recreational fisheries. Determining whether there is a match or mismatch between how data are collected and alternative approaches to fisheries management greatly depends on the human element. Identifying and evaluating alternative management approaches and their intended benefits, and ascertaining the relevant data needed to support them, means tracking the human dimensions of the programs. To provide one example, shifts in management actions result in both expected and unexpected shifts in angler behavior. Anticipating potential shifts in data collection needs to match changes in management requires an understanding of this behavior. This chapter discusses the reasons why these other requirements need to be addressed. In addition, the recommendations that follow address more than just human dimensions; some will strengthen the survey, some will derive additional value from the survey, and some will add to what is done now.

The various surveys currently conducted by NMFS have the potential to provide critical insight to the human dimensions of recreational fisheries on a direct (i.e., during the survey) or indirect (i.e., after the survey) basis. While most of the surveys presently are designed to produce insight to the extent of catch and catch per unit effort (CPUE), it is also possible to gather social and economic data simultaneously or independently as per the requirements of the Magnuson-Stevens Fishery Conservation and Management Act, the national standards therein, the National Environmental Policy Act, and the host of other regulatory requirements addressed by fishery management plans. Currently, the MRFSS collects some sociocultural data (e.g., number of days fishing per year and angler state, zip code, and county of residence), but the focus of the MRFSS and most other NMFS surveys is on catch, harvest, discards, and effort.

Collection of human dimensions information, such as angler attitudes, motivations, management preferences, expenditures, and demographics, can take place onsite during an intercept survey if they serve the objectives of the survey (Green et al., 1991; Pollock et al., 1994). Alternatively, the sampled anglers' information, collected during the creel intercept, can be used to facilitate follow-up or add-on surveys via telephone or mail if, for example, data collection requires more time than is available at the intercept interview (Pollock et al., 1994). This latter approach would be a combination of an onsite fishery-dependent survey with an offsite human dimensions survey. This approach has been used previously in conjunction with MRFSS sampling to collect sociocultural data from anglers on a species-targeted basis in the southeast and northeast United States.

Yet, socioeconomic data on recreational fishing through the MRFSS are collected only rarely; the most recent data collection effort was completed in 2000. In addition to the nationwide valuation and expenditure surveys associated with the MRFSS, several other surveys have been conducted to gather additional information regarding angler behavior and characteristics, such as Northeast Recreational Anglers: Preferences for Fishing and Management Alternatives, the Gulf Reef Survey, and Tackle Retailer Profiles (National Oceanic and Atmospheric Administration, 2000a). In addition, the Large Pelagic Survey collects some socioeconomic information, which is used to estimate the demand for and value of the large pelagic fishery among anglers. But the infrequent, inconsistent timing of these surveys does not provide the ongoing monitoring of the recreational sector that is needed to better inform management decisions.

MANAGEMENT USES FOR DATA

The purpose of evaluation of research efforts is to determine whether agency programs targeting anglers are working and producing the intended benefits. An agency's ability to lead and serve the public depends to a great extent on its ability to continue, modify, or terminate its programs when necessary. An evaluation of the NMFS programs, including its various survey efforts, would be useful to understand angler sentiment in a systematic way and whether intended benefits are being achieved. Other topics would include an evaluation of how open the angler public feels the fishery management process is, how they rank NMFS and the fishery management councils as sources of information and educational materials, and finally, how they rate the effectiveness of the various angler programs. Human dimensions research can be the basis for changing existing agency efforts to be more effective, developing new program elements, or reducing support in favor of alternative efforts. Occasionally, evaluation efforts reveal unanticipated outcomes that agencies should be aware of so they can take appropriate action.

Because of the diversity of angler motivations, the product of recreational fishing is not necessarily the number or size of fish caught but rather anglers' satisfaction level with recreational fishing overall or on the particular day they were intercepted. If NMFS seeks to maximize angler satisfaction as a management goal, they must know something about the importance of various motivations to anglers and the extent to which they are achieved in their fishing experience.

Human dimensions information about marine recreational fishing would include anglers' characteristics, such as age, income, boat ownership, information about their choice of fishing destination (e.g., cost, availability of friends or family, and fishing experience), and how often they fish. These data would be used to form angler profiles for the population and to develop or to provide input data for economic models. These models could be used to evaluate the effects of management policies and to simulate the possible effects of proposed management policies.

The identification and characterization of various stakeholders, including marine anglers, is perhaps where most human dimensions research has focused previously. Angler profiles provide managers with the most basic of information on their clientele. These profiles have seen a shift away from simple means and other measures of central tendency for the angler population (i.e., the "average angler" approach) to identifying groups using market segmentation techniques. This involves partitioning anglers into groups with similar characteristics (e.g., coastal residents and nonresidents, tournament and non-tournament participants, private boat and for-hire anglers); thus, the anglers within each group are likely to be more similar in fishing behavior and attitudes. Once the angler population is partitioned to form subgroups of managerial concern, profiles can cross-tabulate groups by demographic characteristics; participation frequency measures; motivations for participation; attitudes, beliefs, and knowledge; management expectations and preferences; and satisfaction measures.

There are various measures of onsite angler participation including fishing frequency, location, angler expenditures, and mode of fishing. Fishing frequency (or avidity as it is often described) is a measure of fishing experience along with number of years of previous participation. (In the MRFSS, number of days fishing in the past two months is used.) Fishing participation begins with a point of origin (location of primary residence) and ends with a location where the angler was intercepted, as well as mode of fishing that day (e.g., shore, for-hire sector, private and rental boat). See Chapter 3 for a more detailed discussion. Additionally, some anglers participate in fishing tournaments, and some do not; some belong to fishing clubs and organizations, and some do not. Other useful participation measures include self-perceived assessments of fishing skill, as well as fishing-related knowledge and an assessment of how important recreational fishing is to them compared to their other outdoor recreational activities.

Understanding angler preferences for various management measures prior to implementation is important to understanding compliance probabilities. Previous approaches for understanding angler preferences for various management measures have depended mostly on opinion measurement techniques, whereby it was not possible for anglers to consider fully the tradeoffs involved, an approach that yields many socially acceptable "yes" answers.

In contrast, stated preference models make use of hypothetical scenarios to derive individuals' preferences for various management components (Äas et al., 2000). This approach assumes that complex decisions are based not on one factor or criterion but on several considered jointly. Results allow managers to understand how anglers combined their preferences for various management measures under consideration and the relative influence of each management attribute (Louviere et al., 2000). Using a mail questionnaire format, Hicks (2002), for example, identifies anglers' stated preferences for summer flounder regulatory alternatives as an add-on survey to the MRFSS in the northeastern United States.

There are many important human dimensions questions today that involve change over time and require longitudinal study designs. They include questions about trends for anglers joining clubs and associations to gain a voice in management, about rates of participation in fishing tournaments, and about annual fishing frequency. Also, to what extent are attitudes toward catch and release changing over time? These questions will require longitudinal measures using the same questions over time with the saltwater angler population or with angler panel studies.

ECONOMIC DATA AND MODELS

As noted above, marine recreational fishing data often include an economic component—indeed, more broadly, there is sociocultural information as well. The economic data include information on the characteristics of anglers (e.g., age, income, boat ownership), trip choice, expenditures, and other related information. These data are used to form profiles of the angler population and to develop economic models for analyzing fisheries policies. The data may be gathered in an expanded version of an existing recreational fishing survey focused on catch and species information, as a follow-up to such a survey at a later time (possibly using a different survey mode), or as an independent survey with a new sample.

While there are a variety of economic models that use recreational fishing data, the two most common are economic valuation models and economic impact models. Other applications, such as bioeconomic models, participation rate studies, marketing studies, and recreation supply studies, for the most part, use similar data. The purpose of the valuation and impact models and their data requirements is discussed below. There is also a specific set of recommendations at the end of this chapter for accommodating economic data in recreational fishing surveys.

Economic valuation models consider the behavior of anglers and, as their name suggests, are used to value fishery resources. For example, they may be used in cost–benefit analyses of fisheries and environmental regulations, in damage assessment, and in setting management priorities. The following are some examples of the types of questions valuation models can address:

- What is the economic value of an increased catch rate for a specific fish species or group of species in a given region?
- What is the short-term economic loss of closing a recreational fishery? What is the long-term gain if a fishery recovers?
- What is the economic loss to anglers due to a consumption advisory?
- What is the economic value of improved coastal access for anglers?
- What are the relative values of additional recreational versus commercial catch?

Also, valuation models are used to predict the response of anglers to regulatory changes, which in turn may be useful for management and planning at local and regional levels. The following are some examples of behavioral response questions that may be considered:

- How will anglers respond if a recreational fishery is closed? Will they fish fewer days in total? Target another species of fish? Delay fishing until the fishery reopens? Substitute another form of outdoor recreation?
- How will anglers respond to a consumption advisory?

- How will anglers respond to catch limits or other restrictions on a fishery? Will they change fishing effort, targeted species, chosen fishing site, or mode of fishing?

Valuation models come in two basic forms: revealed preference and stated preference. Revealed preference models infer values from fishing choices actually made by anglers. Anglers implicitly reveal economic values in the sites they choose, the species they target, the modes they select, and the frequency with which they fish. Revealed preference models are designed to measure implicit values for fishing using data on observed choices. Stated preference models, on the other hand, ask individuals to state their values directly in a survey. The former has the advantage of being based on actual behavior; the latter has more flexibility in the scenarios it can consider. Additionally, the models may be combined.

There are numerous revealed preference models of recreational fishing, but the travel-cost random utility model is the standard. McFadden's (2001) Nobel lecture provides an excellent exposition on random utility theory. Parsons (2003) presents a review of the model as it is used in recreation demand. There are numerous applications to marine recreational fishing (e.g., Huppert, 1989; McConnell et al., 1994; Gautam and Steinback, 1998; Haab et al., 2001).

The travel-cost random utility model uses data on actual trips to fishing sites to model where anglers fish, how often they fish in a season, what fish they target, what mode of fishing they use, and how long they stay onsite. The model is designed so that choices depend on the characteristics of the site and the characteristics of the angler. A model may include one or more of these choices. The model predicts outcomes in a probabilistic form. For example, the probability that an angler visits a site might increase with the quality of fishing at the site, ease of access to the water, amenities at the site, proximity to the angler's home, and the angler's years of fishing experience.

The basic data requirements for estimating a travel-cost random utility model using marine recreational fishing data are the following:

- A probability sample of anglers and potential anglers
- The location of each angler's residence
- The characteristics of anglers believed to influence site choice, mode choice, and species targeted
- The location and a clear definition of each fishing site

- The characteristics of sites believed to influence anglers' choice of sites over a season
- Each angler's choice of sites visited, species targeted, and mode used over given season

A probability sample of anglers is needed to make accurate inference on the complete fishing population. This is critical if the economic analysis is to be used in policy—the values and behavioral results should be representative of the population. The location of the angler's residence and the location of the fishing sites are necessary in the calculation of travel cost by each angler to each site. The model bears the title "travel cost" for a good reason—travel cost is invariably an excellent predictor of site choice, and it is the factor that anglers use to trade off money and time for better sites and better fishing. Ultimately, travel cost is the way values for sites and their attributes are inferred. Site and angler characteristics are also needed in the behavioral models as predictors to create a realistic model of how anglers make decisions. Finally, actual choices of the site visited, species targets, and mode used are needed as the dependent variables to be modeled. The data should distinguish between primary purpose and side trips, which can be done easily as part of the survey design. In the analysis, the side trips can be handled by changing the origin of the trip. In many instances, the side trips are set aside entirely. In general, clear definition of trip length, purpose, and activities will make for a richer data set from which better analysis can be done.

The data requirements then are twofold: angler-specific characteristic data and site-specific characteristic data. The former are gathered in a recreational fishing survey, and the latter usually are gathered separately as an inventory of relevant sites. While it is difficult to make generalizations about the angler-specific data required for estimating a travel-cost random utility model, Table 5.1 provides some guidance for marine recreational fishing surveys hoping to accommodate economic analysis; the list also includes information that would be useful in a variety of sociocultural analyses as well. This list is not meant to be exhaustive, nor is it meant to be a necessary list for doing an analysis. Rather, it is meant to be representative—one that incorporates most of the important characteristics that show up in contemporary analyses.

TABLE 5.1 Angler-Specific Characteristic Data to Accommodate Economic Valuation Models of Marine Recreational Fishing

Data Type	Characteristics
Angler-specific	Location of residence (city and zip code)
	Demographics
	Gender
	Family size
	Years of fishing experience
	Age
	Boat ownership (yes/no and size)
	Location of vacation home
	Favorite and preferred species of fish
	Income
	Occupation
	Employment status (e.g., retired)
	Education level
Trip-specific	Destination (launch point and at-sea location)
	Mode (e.g., shore, for-hire sector, private boat)
	Species targeted
	Species caught
	Time onsite
	Day trip (hours onsite)
	Overnight trip (days away from home)
	Expense of bait, tackle, and other supplies
Stated preference data	Behavioral response questions to support management needs

Trip and catch recall is always an issue in recreational fishing surveys. At one extreme, the survey may ask for detail only about the last trip taken; at the other, it may ask for detail on all trips in a season. Somewhere in between these extremes is preferred: trips in the last month or two. Since revealed and stated models are often combined, a data element called "stated preference data" has been included. As discussed earlier, these are data elements in which individuals are asked to respond to hypothetical questions, such as changes in fishing laws and

TABLE 5.2 Site-Specific Characteristic Data to Accommodate Economic Valuation Models of Marine Recreational Fishing

Quality of fishing—catch, abundance, and success rate
Size of site (length of coastline)
Type of water body (e.g., ocean, bay, river)
Number of boat ramps and lifts
Population density at site (e.g., urban, rural)
Type of site (e.g., pier, beach)
Availability of facilities (e.g., bathroom, food, bait shops, boating, gas/repair, camping)
Level of regulatory control
Availability of natural cover
Availability of parking

catch rates. To model participation in marine recreational fishing, it is important that data be gathered on those who choose to fish as well as those who choose not to fish. Also, if the data are gathered as a panel (so anglers respond to a survey that reoccurs every two months over one year), there is a single initial collection of the demographic data. Thereafter, each angler is asked only about trips in the preceding two months, which shortens the later surveys.

Site-characteristic data, which are gathered separately, are an inventory of characteristics that can be compiled using existing state agency data, field visits, tourist guides, and fishing guides. The catch data may be obtained separately from a creel survey or in the angler survey, but some calibration using both is preferred. Again, it is difficult to make generalizations about the site-specific data required for estimating a travel-cost random utility model, but Table 5.2 provides some guidance.

The other economic valuation models mentioned above use stated preference analyses. In these analyses, values are not inferred from actual behavior. Instead, analysts pose hypothetical trip or valuation (willingness to pay) questions to anglers. As noted above, this approach has more flexibility but is less conducive to a general template or guide for data collection since the valuation questions are likely to vary with region and to be specific to policy needs. In a national study, there may be some merit in considering a rotating set of stated preference questions for loss of a specific site or sites and for change in the catch of specific fish species. A time series on values such as these may be useful for policy makers. In addition, state preference questions can be adjusted regionally and over time to meet specific policy needs.

Mitchell and Carson (1989) provide an excellent background on the application of surveys to value public goods using stated preference methods. For some applications to marine recreational fisheries, see a study of summer flounder (Hicks, 2002), an analysis of fisheries management options (Oh et al., 2005), a study of salmon and striped bass in San Francisco Bay (Cameron and Huppert, 1989), and a study of saltwater and freshwater fisheries in Washington State (Layton et al., 1999).

The other analyses conducted with marine recreational fishing data are studies using economic impact models. These are sometimes called input–output models and attempt to track the impact of regulatory changes through a local or regional economy. They are used to answer questions, such as the following:

- What is the local and regional economic impact on different user groups and industries of a catch limit or area restriction for a specific fish species or group of species?
- What is the impact of improved access to a site or of a new marina at a given site?
- What is the impact of the complete collapse of a recreational fishery?

Consider the collapse of a fishery. The local economy would be affected through a drop in the sales of bait and fishing equipment, sales of gasoline, visits to restaurants, visits to nearby attractions, stays at hotels, and so forth. These declines in economic activity, in turn, lead to decline in demand for other goods and services by the producers of these goods and services. Therefore, in turn, summer employment may decline, groceries sales may fall, and so on. In this way, the impacts of the fishery ripple through the economy. The shortcoming of these models is that they ignore the impacts outside the local economy and region. A declining fishery, for example, may leave an inland angler at home to spend his money with a positive (and ignored) impact there. Still, local and regional regulators often demand impact studies.

Unlike the travel-cost random utility model, impact studies rarely, if ever, develop models specifically to study fisheries impacts. Instead, existing impact or input–output models are used. The most widely known model is IMPLAN (Impact Analysis for Planning).[1] A number of trip-specific expenditure variables (e.g., transit costs to site, food, bait and tackle, launch and boat fees, fuel and rental costs, lodging) and durables data (e.g., value of boat, electronics, and rods and reels) could be targeted to accommodate a typical impact study. Like stated preference studies, data collection efforts for impact analyses could be flexible enough to be adjusted regionally and over time to meet specific policy needs.

CONCLUSIONS AND RECOMMENDATIONS

The current MRFSS was not designed with human dimensions data in mind. Much of the data is collected through add-on surveys that suffer from many of the same design problems associated with collection of catch and effort data. There is the potential to collect high-quality human dimensions data, but this has never been a traditional component of the MRFSS and other NMFS surveys. Despite the numerous important human dimensions questions identified earlier in this chapter, a human dimensions perspective on catch and effort has not been a priority of NMFS data collection efforts. However, with the amount of money currently allocated to support the MRFSS and the amount that might be necessary to support a redesigned MRFSS, an integrated approach to fisheries management and the collection of requisite data is essential.

With respect to the economic models, **add-on surveys for human dimensions should be continued but in a more focused way than is done currently to target specific management needs and to supplement the national data as needed.** Traditional add-ons are “choice-based” onsite samples (i.e., access-point intercept surveys for CPUE) that make extrapolation to the population of users unreliable. Add-on surveys that build on the samples to develop effort estimates (i.e., offsite random digit dialing surveys) provide a better sampling frame for the choice component of the data. Unfortunately, these data have been constrained to the population of anglers within 25 miles of the coast, which severely limits the ability of the models to make inference about the relevant population of anglers. Also, surveys that gather biological and economic data simultaneously place a large burden on respondents, which may

---

[1] Refer to Dietzenbacher and Lahr (2004) for more information on the structure, theory, and history of impact models. For some examples of impact studies, see Steinback (1999) and Bohnsack et al. (2002).

lower the quality of both data sets through lower response rates and interviewer fatigue. Simultaneous surveys also can remove flexibility in timing, design, and sampling, which may vary in the economic and biological components. Finally, the data set (or inventory) of marine recreational fishing sites and their characteristics lacks some needed data. For this reason, analyses often use limited site characteristics in the models (such as catch and travel cost only), collect the site data independently, focus on more limited policy needs, and estimate less defensible models. Economic valuation studies, marketing studies, business interests, and even data collection efforts regarding catch would benefit from a carefully designed data set on marine recreational fishing sites that is updated regularly for accuracy. **If the number of marine fishing trips increases, it is likely that additional fishing access sites will be developed. In addition, social and environmental changes (e.g., changes in the distribution and numbers of people, a major hurricane) also can affect the availability and use of access sites. To ensure adequate coverage of the recreational fishery, a periodic updating of lists and descriptions of fishing locations and access sites is needed.**

An independent national trip and expenditure survey should be developed to support economic valuation studies, impact analyses, and other social and attitudinal studies. This survey should follow these guidelines:

- Use a random sample of anglers from the national registry or license frame (see earlier recommendation) and collect the data independent of the catch and effort survey
- Gather data on anglers and their choices (see Table 5.1 as a guide)
- Conduct the survey continuously and as an annual panel for trip data, and every five years for expenditure data
- Use multiple survey modes—mail, phone, internet, in-person—to gather data
- Target response to exceed 50 percent
- Target annual sample size of respondents to be at least 1,000 anglers in each fishery council region
- Include behavioral response questions for verification and to meet specific policy needs

The design of the national human dimensions survey should be independent of the MRFSS catch and effort survey to better align the surveys to their respective purposes, to give adequate flexibility on both the economic and biological sides, and to reduce respondent burden. However, the sites sampled should be the same for the national economic and the MRFSS surveys, as described below. The survey should be conducted throughout the year to develop good seasonal profiles. Survey questions should ask about trips no more than two months prior to avoid recall problems and to keep the survey short to avoid interviewer fatigue. The questions on expenditures should focus on the last trip only for the same reasons.

If time and other resource constraints are limiting, less frequent sampling (every other year or every third year, for example) would be preferred to a lower sample size, lower response rate, and "convenient" sampling strategies tied to onsite surveys. High response rate and probability sampling should be high priorities because they maintain the quality of the survey. If the survey must be conducted as an add-on, it should be part of the effort survey, not the onsite CPUE survey. Also, information on the angler's hometown and destination of each trip is essential for conducting the valuation models. The other data elements in Table 5.1 are of next importance, and expenditure data would have lowest priority. In the absence of a national registry or license frame, the same data outlined above using a sampling frame that covers the entire country but stratified to oversample coastal counties using a combined telephone–mail–internet survey would be the best alternative.

The national database on marine recreational fishing sites and their characteristics should be enhanced to support social, economic, and other human dimensions analyses. The database should:

- Geo-code and define sites at levels as fine as possible
- Gather data on site characteristics (see Table 5.2 as a guide)
- Use multiple resources, such as field visits, travel guides, and state agency data files, to gather the data
- Be updated periodically
- Coordinate with other surveys on catch and species information
- Include historic trip counts and fish catch
- Develop an "on-the-water" site inventory (i.e., document where people fish on the water)

6

Program Management and Support

As noted in earlier chapters, collection and processing of timely and scientifically credible data for recreational fisheries is extremely challenging because of complex survey designs and measurement issues; furthermore, these challenges are evolving over time. Many shortcomings in the current statistical system for marine recreational fisheries have been identified in this report and in earlier reviews (e.g., National Research Council, 2000, 2004).

Sampling and measurement methods used by the Marine Recreational Fisheries Statistics Survey (MRFSS) are widely questioned, and several states have opted out of the current national system, leading to either a duplication of effort or difficulties in standardizing methodologies. Estimation methods do not take full advantage of modern statistical techniques, particularly the use of auxiliary information.

RESEARCH NEEDS AND PROBLEM SOLVING

Currently, a great number of biases exist, and assumptions are made because the sampling methodology is often inadequate to allow for accurate data analysis. Therefore, research is needed to determine how best to reduce biases and assumptions. Since data collection is not supervised by an overarching group, research also needs to be done to improve the accuracy of data collection.

Previous chapters outline the issues and concerns with the current survey methods. A number of ideas have been suggested for circumventing or ameliorating these issues, including establishing a better sampling frame, tracking individuals through a panel survey, making use of auxiliary information where appropriate, and conducting experiments to identify and account for biases and to estimate influential factors that otherwise cannot be determined (e.g., the estimation of release mortality rates). The hiring of additional fulltime statisticians would increase the expertise within the statistical offices and would improve data analysis and therefore fisheries management decision making.

A NATIONAL STATISTICAL PROGRAM

The development of a national statistical program for marine recreational fisheries data could be used to eliminate survey effort redundancies. Such a program might consist of a federal agency, regional and state offices, and an independent research group (IRG). The first two of these components already exist, and their specific responsibilities could be delineated to minimize duplication. Briefly, the federal agency provides a nationally consistent statistical system, while the regional offices offer on-the-ground expertise and implementation capability. The third program component, which would be new, is necessary to ensure sufficient staffing and other resources to implement changes, to develop state-of-the-art design and estimation systems, and to anticipate and adapt to evolving challenges. It is unlikely that a federal agency with so many other responsibilities would be able to deal with these issues as well. Since methods should not be reinvented region by region, regional offices also should not have this responsibility.

Independent Research Group

An IRG with recognized expertise in theory and methods of survey design; survey operations; human dimensions; and statistical estimation techniques, including modern survey regression techniques, time series analysis, and small area estimation, could revolutionize recreational fisheries data extrapolation and understanding. The IRG should have a proven record of research publication in the peer-reviewed literature, and ideally, the IRG would have ready access to additional expertise in fields like remote sensing, landscape ecology, cognitive science, computer science, and economics. Such a structure would provide the flexibility to anticipate and adapt to evolving challenges. Though a small number of private research organizations might meet these constraints, the IRG most likely would be a unit within a research university or consortium of universities. Such academic environments tend to produce research that is innovative and of high quality because much of it is ultimately subjected to rigorous peer review. Additionally, involvement of a research group outside the federal agency should help to increase stakeholder confidence in the statistical system.

This IRG would not be responsible for data collection or dissemination and may or may not be responsible for online quality assurance and production of estimates, but it should be in a position to interact closely with other program components in these operations. This interaction ensures that methods developed by the IRG are operationally feasible and modifiable in real time, if necessary. Other responsibilities of the IRG might include some or all of the following:

- Collaboration with the other program components on the establishment of nationally consistent standards for design of marine recreational fisheries surveys and for data and metadata
- Production of detailed sampling designs and data collection protocols
- Thorough evaluation of existing sampling designs, data collection protocols, and estimation techniques
- Technical documentation of all design, edit, imputation, and estimation procedures and assistance to other program components in explaining these methods to stakeholders
- Directed research on design and analysis to address specific data needs and known challenges, including continuing evaluation of developments in survey theory and methods that may have application to recreational fisheries
- Basic research on new and innovative approaches to marine recreational fisheries surveys and to the analysis of such survey data
- Outreach to the scientific community through conference presentations and peer-reviewed publications and to regional offices on best statistical practices for marine recreational fisheries

A model for such an IRG in a university setting is the statistical unit supported by the large-scale, long-term cooperative agreement between the U.S. Department of Agriculture's Natural Resources Conservation Service (NRCS) and the Iowa State University Statistical Laboratory. NRCS and the statistical unit cooperate in producing the National

Resources Inventory (NRI), which is a stratified two-stage area sample that provides information to support agricultural and environmental policy development and program implementation (Nusser and Goebel, 1997; Breidt, 2002). In its current form, NRI is an annual rotating panel survey designed to assess conditions and trends for soil, water, and related natural resources on nonfederal lands of the United States (Breidt and Fuller, 1999). The survey has evolved over a period of several decades, and before adopting its current annual form, NRI was conducted every five years, assessing conditions on each of approximately 800,000 sample points throughout the United States (U.S. Department of Agriculture, 1999). Sample points are located in all counties and parishes of the 50 states and in Puerto Rico, the Virgin Islands, the District of Columbia, and selected portions of the Pacific Basin (Breidt, 2002). In addition to providing monitoring information, NRI is used as a frame for special studies conducted outside the normal sample rotation (e.g., wetland changes, soil quality).

The Iowa State University Statistical Laboratory provides technical expertise on design and estimation for NRI. Specialized methodologies have been developed at Iowa State for activities, such as editing, imputation, small area estimation, and variance estimation. The relationship between Iowa State and NRCS goes back many years, with contracts renegotiated annually over most of that time. Most recently, they moved to a five-year contract obtained in a competitive bidding process.

The current five-year contract is about $10 million and is for design and estimation with specific products, including an imputed dataset with appropriate weights. The funding does cover design of data collection protocols, including extensive interaction with three federal data collection centers, but does not cover actual data collection. The contract supports roughly 1.5 faculty, 12 professional and scientific staff (e.g., programmers, geographic information system [GIS] specialists, database managers), 5 graduate research assistants, and parts of several administrative staff (e.g., accountants, secretaries). Four NRCS employees are located at Iowa State to allow them to work with Iowa State staff and participate in NRI design and estimation development, but these NRCS employees are not supported by the contract. Because the contract is in an academic department at a public research university, there is an expectation that theoretical and methodological research arising from the contract is broadly disseminated through scientific conferences and peer-reviewed publications. Some of the contract funding is used explicitly to support this dissemination.

Two Science and Technology to Achieve Results (STAR) research assistance agreements, which are awarded by the Environmental Protection Agency (EPA) to faculty members in the Departments of Statistics at Colorado State University and Oregon State University, provide another model for an IRG in a university consortium that supports an official statistical program. Additional key personnel are at other agencies and universities. The two agreements focus on statistical issues arising in the monitoring of aquatic resources, including lakes, streams, and estuaries, with particular emphasis on the EPA's Environmental Monitoring and Assessment Program and related programs. The agreement primarily funded at Colorado State University is the Space-Time Aquatic Resources Modeling and Analysis Program (STARMAP) (Colorado State University, 2005), and the agreement primarily funded at Oregon State University is Designs and Models for Aquatic Resource Surveys (DAMARS) (Oregon State University, 2005). STARMAP is somewhat more focused on estimation issues and DAMARS more on design issues, but there is considerable overlap in the research programs, particularly in small area estimation.

Unlike the Iowa State/NRI agreement, these STAR agreements do not have specific deliverables like a weighted dataset. Instead, the agreements primarily are aimed at pure methodological research. However, an explicit component of the agreements is outreach to federal, state, and tribal agencies responsible for aquatic resources monitoring. This outreach involves helping these agencies to implement the new statistical methodologies developed under the STAR agreements. There is considerable interaction with EPA scientists, primarily through short-term visits. Each of the agreements is a $3 million, four-year contract supporting faculty, graduate research assistants, and post-doctoral research associates. The two STAR agreements are tied to one another through subcontracts, joint annual meetings, and a common scientific advisory committee.

Federal Agency

The role of a federal agency in a national statistical program can vary from complete responsibility to coordination and oversight. Due to the complexities of data collection, processing, and dissemination and to the different scales of observed data, it is unlikely that a federal agency could oversee and meet the needs of all parties adequately. A federal agency is, however, ideally situated for ensuring a nationally consistent statistical system and for setting standards to be coordinated among the different offices. Specific responsibilities of the federal agency might include some or all of the following:

- Management of the overall program to ensure proper collaboration between the IRG and regional and state offices
- Coordination with states or regional offices, including regional fishery management councils, on problems of mutual interest, such as creation and maintenance of a national list frame of marine anglers for efficient sampling and outreach programs
- Collaboration with the other program components on establishment of nationally consistent standards for design of marine recreational fisheries surveys and for data and metadata (possibly including training of access site data collectors to ensure national consistency)
- Cooperation with other program components on quality assurance and quality control for collected data
- Cooperation with the IRG in the processing of data and production of estimates
- Maintenance of a central data warehouse for marine recreational fisheries and development of appropriate dissemination tools
- Outreach to national-level stakeholders, including organizing data product reviews by user groups, providing information to national media, and providing education to policy makers
- Solicitation of feedback on data needs from user groups, such as scientists, policy makers, and managers
- Collection of information from scientists, managers, and data collectors on evolving challenges in marine recreational fisheries surveys
- Communication of data needs and evolving challenges to the IRG
- Sponsorship of directed research on design and analysis to address specific data needs and known challenges
- Sponsorship of basic research on new and innovative approaches to marine recreational fisheries surveys and to the analysis of such survey data

One final responsibility that may fall to the federal agency is solicitation of formal scientific reviews of the entire statistical program for marine recreational fisheries. Such reviews could be done on an ad hoc basis possibly through standing committees. Models for such committees include those maintained by the American Statistical Association (ASA). These standing advisory committees meet periodically to advise federal statistical agencies (American Statistical Association, 2006a). Current examples include the Census Advisory Committee, the Committee on Law and Justice Statistics (advisory panel for the Bureau of Justice Statistics, among other responsibilities), and the Committee on Energy Statistics (advisory panel for the Energy Information Administration [EIA]).

The latter committee is a good example of an ASA advisory committee. EIA is an independent statistical agency within the U.S. Department of Energy. The ASA Committee on Energy Statistics is supported by a line item in EIA's annual budget. This support is administered by ASA. The committee meets twice a year and advises EIA on a broad range of issues, not only on data gathering, data quality, and modeling, but also on strategic planning and stewardship of EIA's reputation in the scientific community. At each meeting, EIA updates the committee on its use of the committee's advice. The committee consists of a chair plus 12 members, including a balance of mathematical statisticians, survey methodologists, economists, energy modelers, and policy analysts. The committee's composition is dictated by the broad range of issues on which EIA needs advice: data collection, data presentation, energy modeling, economics, and policy analysis. Committee members serve rotating three-year terms with an option to renew for a second three-year term at the discretion of ASA and EIA.

Regional and State Offices

As opposed to the previous duties, maintaining access site list frames for catch per unit effort estimation would be accomplished at the regional or state level. Since local expertise on topics like geography, fishing modes, and species variation is essential, data collection is best done at a local level. Additional benefits to having state or local personnel conduct the intercept surveys include better training, the formation of an interviewer–angler relationship, and a larger sample size. These benefits would come with higher costs, but survey reliability, data quality, analysis, and credibility would increase. Even with outstanding local expertise, access site list frames have a number of potentially serious deficiencies as outlined in earlier chapters and should, where possible, be supplemented with area samples or other dual-frame techniques to get at catch per unit effort for anglers not accessing the water from listed public access points. The IRG could assist in designing appropriate supplementary samples. Other responsibilities of the regional offices might include some or all of the following:

- Coordination with the federal agency on problems of mutual interest, such as creation and maintenance of a national list frame of marine anglers
- Collaboration with the other program components on the establishment of nationally consistent standards for design of marine recreational fisheries surveys and for data and metadata
- Collection of data, primarily from angler intercepts at access sites, which ideally would be done with a dedicated staff of state employees to ensure quality and continuity in the data
- Cooperation with other program components on quality assurance and quality control for collected data
- Communication of data needs and evolving challenges to a federal agency and the IRG
- Outreach to local user groups, including communication with individual anglers at access sites

INTERIM SOURCES OF PROGRAM SUPPORT

Establishment of the outlined three-component statistical system for marine recreational fisheries will take time and resources. In the interim, other sources of much-needed technical support should be considered. Many federal agencies directly support relevant statistical research on a short-term basis (one to three years) through contracts and grants, such as EPA STAR grants, Research Joint Venture Agreements from the U.S. Forest Service, and cooperative agreements with the U.S. Census Bureau or the U.S. Bureau of Labor Statistics. These may be based on broadly advertised requests for proposals or may be directed by the agency to a particular researcher or group.

Another mechanism by which federal agencies support academic researchers doing relevant research on survey methods is through the Methodology, Measurement, and Statistics Program (MMS) in the Division for Social, Behavioral, and Economic Research of the National Science Foundation (NSF). MMS works in collaboration with a consortium of federal statistical agencies represented by the Interagency

Council on Statistical Policy and the Federal Committee on Statistical Methodology. (The National Oceanic and Atmospheric Administration is considered a federal statistical agency and is a member of the Federal Committee on Statistical Methodology.) MMS can fund relevant research in two ways: (1) directly through a call for proposals on topics of interest to the federal agencies or (2) indirectly through scanning proposals to other NSF programs (notably pertaining to statistics and probability in the Division of Mathematical Sciences) and jointly funding those of interest to the federal agencies.

According to a recent call for proposals, MMS invited research proposals

> …that further the development of new and innovative approaches to surveys and to the analysis of survey data… Although proposals submitted in response to this solicitation may address any aspect of survey methodology, priority will be given to basic research proposals that are interdisciplinary in nature, have broad implications for the field in general, and have the greatest potential for creating fundamental knowledge of value to the Federal Statistical System. Because methodological problems often require knowledge and expertise from multiple disciplines, collaborations are especially encouraged among the relevant sciences, including the social sciences, linguistics, cognitive science, statistics, computer science, and economics. (National Science Foundation, 2005)

Such a call could be directly relevant to the development of survey methods for marine recreational fisheries.

One other useful means of providing interim technical support to the statistical program is to bring short-term academic visitors into the federal agency through a formal fellowship program. ASA joins with NSF and various federal statistical agencies in supporting the ASA/NSF/Federal Statistics Fellowship program (American Statistical Association, 2006b). The program brings academic researchers into the federal agencies for up to one year for the purpose of fostering collaborative research efforts on methodology relevant to the agencies. The roles of ASA and NSF in the fellowship program include advertising the fellowships to the relevant audience, forming the committee that reviews the applicants, and lending prestige to the fellowship awards.

Agencies that have participated in the ASA and NSF fellowship program have included the Bureau of Labor Statistics, the Census Bureau, the Bureau of Economic Analysis, and the National Center for Education Statistics. Similar research programs have been jointly sponsored (without NSF) by ASA and the National Center for Health Statistics, the National Agricultural Statistics Service, and EIA.

CONCLUSIONS AND RECOMMENDATIONS

A large number of complex technical issues associated with surveys of marine recreational fishing remain unsolved, and a significant investment in intellectual and technical expertise therefore is needed. Research is required to determine how best to reduce biases and assumptions and to improve the accuracy of data collection. To address these needs additional fulltime statisticians should be hired by the MRFSS. **A research group of statisticians should design new analyses based on current developments in sampling theory** (as outlined in Chapter 3). These examinations should include experimentation, such as specific sampling of activities like night fishing or fishing from private property, whose current underrepresentation in the MRFSS sampling has the potential to create bias.

A greater degree of coordination between federal, state, and other survey programs is necessary to achieve the national perspective on marine recreational fisheries that is needed. The committee recommends the development of a national statistical program that might consist of three components: a federal agency, regional and state offices, and an IRG. **A permanent and independent research group should be established and funded to continuously evaluate the statistical design and adequacy of recreational fishery surveys and to guide necessary modifications or new initiatives. Human dimensions expertise should be included as well. The recommended changes to the design and operation of the MRFSS and its continued development and operation will require additional funding above current levels.** In the interim, the committee recommends that other sources of funds be considered for the technical assistance that is needed immediately for the MRFSS.

7

Communication and Outreach

Ultimately, the value of marine recreational fishing data, whether collected by the Marine Recreational Fisheries Statistics Survey (MRFSS) or any other survey, will be judged by the extent to which it meets the needs of the individuals who use the data and will be trusted by those whose lives are affected by the ways the data are used. This latter group of stakeholders is a varied group including saltwater anglers, other user groups who benefit from use of recreational fishery resources, and commercial fishermen whose benefits from the fishery resource may be influenced by allocations made as a result of available data. Stakeholders rarely understand why fisheries data are collected in certain ways and how the data are analyzed and applied to management decisions. Communication and outreach efforts are essential to foster confidence in the quality of the data among managers, other decision makers, and those who rely on the fishery resources for recreation or for a living.

A primary challenge to be addressed in communication and outreach efforts is the disparity in how data are perceived by various stakeholder groups, including data collectors, data analysts, and the recreational and commercial fishing communities. As described in Chapter 1, the MRFSS was originally designed to characterize the nature of and trends associated with recreational fisheries, particularly in terms of catch, effort, and participation at national and sometimes regional scales. Among the recreational fishing community, there is a widespread lack of support and appreciation for the current MRFSS administered by the National Marine Fisheries Service (NMFS). This lack of support is related, in part, to the evolution of the use of the MRFSS data. Since its inception, the MRFSS data have been applied to other purposes, most notably stock assessment and management decisions for particular species, which were not the original intent of the MRFSS. During that time, the types and numbers of marine recreational fishery stakeholder groups have grown, as has the recognition of the importance of knowing accurately the size of the recreational harvest as a portion of total harvest in many marine fisheries. These trends have contributed to an increased interest in recreational fisheries statistics, not only among anglers but among other stakeholders as well. All of them care about fisheries management decisions, including allocation decisions. Political demands placed on managers to make "good" decisions based on the MRFSS and other data sources, and the scrutiny applied to fisheries statistics and their associated methods, also have increased. In addition, it is difficult for an individual angler to distinguish population trends (e.g., depletion) from fluctuations caused by weather, local fish migration, and other factors that affect catch rate. It also is difficult for an individual angler to understand the cumulative effects of many anglers on the same species.

With all of these competing demands, communication and outreach become even more critical to ensure a shared understanding of the purposes, capacity, and limitations of the data-gathering approach, and consequently, the quality of or limitations on uses of the data. These concerns are relevant for any type of fisheries statistics, whether it is for the MRFSS, some existing MRFSS-like state or regional survey, or even commercial fishery statistics.

A previous National Research Council study notes the following:

> It is important for scientists and managers to improve their communication of the data available and to make such data available to stakeholders more readily and in a user-friendly form. When this is not achieved, a lack of trust develops between those who control access to the data and those who cannot gain access. In many cases, disagreement of fishermen with the results of stock assessments can be traced to [NMFS] not explaining the sources of variability in the data and the uncertainty of the models being used. (National Research Council, 2000)

What should this communication and outreach involve? Fundamentally, communication programs must identify and respond to the information needs of each of the stakeholder groups involved in the marine fishery management milieu. This requires identifying the appropriate stakeholder groups who have particular information needs or information to share, and articulating the information and knowledge each should have or wants to have (communication and outreach goals). It also requires identifying the most appropriate means for exchanging information among the various stakeholder groups, including the recreational fishing community, data collectors, and data analysts and users, taking into account that the understanding of probability and statistics in the general population is not high. For any communication program, it will be important to remember that communication should be an exchange of perspectives and information, not just a one-way presentation of information.

COMMUNICATION GOALS AND STAKEHOLDERS

The importance of clear and ongoing communications between scientists, managers, and fishermen (both anglers and commercial fishermen) has been articulated most clearly in relation to the commercial fishing sector. Jentoft et al. (1998) argues for actively engaging fishermen in fishery management processes to improve management decisions by including experiential knowledge of those involved in the fishery, improving communication among all parties involved, and increasing the consideration given to socioeconomic aspects of the fishery in decision-making processes. Johnston (1992) articulates a range of benefits accrued through partnerships among commercial fishermen, scientists, and managers, including a greater commitment by the fishermen to achieving management successes when they have an active role in designing management strategies, reducing data-gathering costs because fishermen become more cooperative in the data-gathering efforts, and enhancing the credibility of the management process and decisions based on data collected. All of these benefits can lead to greater acceptance of fishing regulations.

Commercial fisheries management has pushed the concept of collaboration further than open, two-way dialogue by experimenting with comanagement schemes that legitimize management decision-making authority for fishermen (e.g., Yandle, 2003). True comanagement has received limited attention for recreational fisheries management (Wilson et al., 2003), but many of the concepts underlying comanagement approaches suggest lessons for at least improving the relationships between fisheries managers, fisheries scientists, and anglers. Attention to communication and outreach may improve relationships among data collectors and analysts, managers, and decision makers who use the data and the private and for-hire sector fishery participants whose lives or livelihoods are affected by the data and their use.

This committee heard from numerous groups and individuals expressing a lack of confidence in the estimates produced by the MRFSS. This credibility gap arises from several sources, including a belief that alternate data sources are more credible; criticism of the temporal, spatial, group, or taxonomic stratification of the intercept sampling; lack of understanding of statistical methodology; and recognition that the sampling frame used for effort estimation suffers from undercoverage. Communication and outreach goals associated with a marine recreational fisheries statistics program might focus on topics related to data collection efforts and to data interpretation and use.

DATA COLLECTION EFFORTS

An overarching goal associated with the data collection process may be to encourage anglers and their representatives to form more positive attitudes toward NMFS so that they will adopt their catch and effort surveys and the results they provide, support management decisions based on these results, and put more trust in the agency. More specific goals may include:

- Enabling anglers to better understand the reasons for using probability samples rather than censuses and the implications for such a sampling method in terms of on-the-ground contacts between anglers and data collectors. The most common criticism made by active anglers and the people that represent them regarding the MRFSS is they have not been intercepted by survey personnel working in support of NMFS. An onsite avidity bias toward those who fish most frequently would suggest they have a greater chance of being intercepted than more casual anglers. Also, more avid anglers are likely to be more skilled than casual anglers, which implies that their catch rates are higher than those of casual anglers. Most anglers are not likely to understand the intricacies and efficiencies of random sampling designs and would question their use compared to a census. Outreach activities should emphasize the process of data collection, as well as results.

- Ensuring that anglers understand the basics of sampling and the importance of a frame from which to draw a sample. Such knowledge may help build support for implementation of a national registry or saltwater fishing license in their state (if they do not already have one). Anglers should have a clear understanding of the link between having a saltwater fishing license, and thus a sampling frame, and improving the quality of recreational fisheries data and stock estimates based on those data.
- Improving the quality (validity) of the data collected by improving the confidence anglers have in data-collection efforts. If anglers are more invested in the survey process, they likely will be more motivated to participate and report accurately and be less likely to complain about a method they helped implement.
- Improving the design of data collection efforts by providing a mechanism for structured feedback from users regarding design characteristics. Although several external reviews of the MRFSS or portions of it have been conducted (Guthrie et al., 1991; Pollock et al., 1994), an internal process of user feedback on evaluation and modification of the design, currently, is not available within the program. Some users of the MRFSS data have initiated dialogue with the MRFSS project managers to address design issues, but there is a need for a more formal and institutionalized feedback process. This committee judges that the lack of such a process may be because there has been no formal re-evaluation of the MRFSS mandate and objectives in relation to current data needs and usage. The rapid evolution of uses and needs for data from recreational fisheries underscores the requirement for such a re-evaluation by the MRFSS managers and communication about that re-evaluation with data collectors, data analysts, and the recreational fishing community.
- Establishing a common knowledge base among anglers, data collectors, and data users. Data collectors, for example, need to understand common names for fish and generally be aware of angler behaviors, gear types used, and species caught. Different modes of fishing (e.g., shore-based, boat) occur at different times and places and often are subject to different regulations. Also, data collectors need to use categories that will meet the requirements of analysts. Concerns have been voiced about the ability and training of those conducting the surveys in support of

NMFS. If anglers are not confident that surveyors have the requisite fish identification skills, then resultant data and efforts to use those data in support of management are not likely to be accepted. If data collectors "speak the same language" as the anglers, confidence in the abilities of the data collectors will be enhanced, and the data collectors should produce higher-quality (more accurate) data. This is particularly important for taxonomic stratification of data. Analysts must be able to employ data with confidence that species designations are accurate and consistently applied in the sampling process. Biological data obtained from the intercept samples must be consistent with the categories used in assessments. This committee heard testimony that suggested that there are inconsistencies in taxonomic segregation of data among different fishing modes. Here again, improved outreach efforts between analysts and data collectors are required, and this includes outreach with the recreational fishery participants as well.

DATA INTERPRETATION AND APPLICATION EFFORTS

Problems associated with marine recreational fisheries statistics are not limited to data collection efforts. As detailed in previous chapters, the MRFSS has been applied to answer questions it was not designed to address. In some instances, misunderstandings have developed because the current use of recreational fisheries data originating with the MRFSS was not anticipated in the design of that program. Current users require data that are more highly resolved—spatially, temporally, and taxonomically—than is currently collected. Mechanisms to modify, amend, or enhance the data collection processes could be identified through better communication and outreach between data analysts, users, and collectors.

In addition to dialogue on design issues, the MRFSS managers also need to advise data users on constraints to some uses, as well as fundamental features of the data collection system. The MRFSS website is information rich and provides general background for the average angler. In addition, the MRFSS personnel conduct regular meetings with users to review results of sampling waves. However, our review identified a number of areas where users extracted sections of data histories but were unaware of the data characteristics, the methods of compilation, or the fundamental nature of sampling versus census estimation. These observations indicate that while the program has undertaken some outreach activities with users, misconceptions and lack of clarity on data characteristics continue to exist. Further, the users' lack of knowledge of and involvement in the design basis of the survey clearly has created some lack of trust in the underlying data presentations. Considerably greater outreach effort appears necessary, although this distrust may not be overcome completely.

Communication and outreach goals focusing on improving data interpretation and application may include the following:

- Ensuring that stock assessment scientists, fisheries managers, and other decision makers are aware of the limitations and inherent biases of marine recreational fisheries statistics related to survey design and approach. Issues that assessment scientists and decision makers should be aware of include the lack of continuity in intercept samplers, differences in sampling methods applied to different modes of fishing (e.g., independent anglers, guided anglers, shore-based anglers), lack of incorporation of design elements in the estimation process (e.g., weighting of spatial or temporal sampling strata), differences in frequency and distribution of fishing trips due to local topography and climate (e.g., rocky shorelines with rough seas in the Pacific Northwest lead to fewer access points than sandy shores and calmer waters in the southeastern United States), and the lack of consistency or accuracy in species designation among fishing or sampling modes. Scientists using marine recreational fisheries data may assume that their statistical properties are known and estimable when in fact they may not be. Resolution of this difficulty can occur only through a detailed outreach process between data collectors and data analysts.
- Facilitating the evolution of the survey system. While one might consider limiting the use of data to the purposes for which they were designed initially, data needs are likely to continue to expand. Communication channels and outreach can be used to identify the growing needs of fishery analysts and other data user groups in the hope that the system can continue to evolve to meet new needs and expectations without circumventing present demands.

COMMUNICATION AND OUTREACH APPROACHES

As one recreational fishing participant commented to this committee regarding the value of a pair of meetings the MRFSS data collectors conducted with recreational fishing participants: "At the first meeting, the MRFSS staff talked at us; at the second meeting, they talked with us." This statement speaks volumes—the key to successful communication and outreach is to talk "with" each other, not "at" each other. There is no question that personal relationships matter, as does creating an atmosphere for honest dialogue and exchange of ideas, rather than one-way information flow.

Communication and outreach to achieve many of the goals noted above can fall under the broad rubric of "public participation" approaches—focusing on how citizens (e.g., anglers, head boat and charter boat operators) can participate more actively and in a more informed manner in management and decision-making processes related to marine recreational fisheries, including data-gathering and data-use efforts. Many public participation concepts developed for other natural resource contexts are appropriate for marine recreational fisheries.

Institutionalize the Importance of Outreach and Communication

Involvement with and leadership for outreach and communication should become an expectation for key individuals at all levels of the marine recreational fishing statistics effort. This includes data collectors within NMFS, as well as data analysts, stock assessors, and decision makers. Individuals' performance plans for their jobs should require the National Oceanic and Atmospheric Administration (NOAA) managers, technicians, and scientists to get out of their offices and interact with anglers and the nongovernmental organizations that represent them; adequate funding should be provided to support such activities. Work hour rules may need to be revised—anglers typically have regular jobs as well and may only be available during evening or weekend hours. Thus, flexible schedules are needed for the NOAA staff who are asked to interact regularly with those in the recreational fishing community. Good relationships between anglers and data gatherers mean that data gatherers must be well-trained, informed, and able to relate to anglers; additional training beyond the biological basis for their jobs will be required. Strategic plans prepared by NMFS should include specific activity targets and outcomes focused on outreach and communication with a variety of audiences and stakeholder groups.

Conduct Regular Regional Workshops

Increasing displeasure among stakeholder representatives over the use of the MRFSS or Large Pelagic Survey data to enforce recreational quotas led to a 2002 workshop, which was convened in San Diego by NMFS and the three fishery management commissions. The workshop involved stakeholder representatives, academics, and various agency personnel and sought to indicate the data types and data collection system appropriate for quota management. Such workshops, when conducted on a regular basis, provide a forum for stakeholders and agency personnel to interact, build relationships, and directly address questions of concern. Given the time and planning that such workshops will require in order to develop relationships and explore topics in depth, it may not be possible to add these as one agenda item among many in regional council or commission meetings; stand-alone meetings may be required.

Engage Anglers in Partnership with Scientists

Engaging anglers in partnership with scientists to collect data to inform stock assessment or other recreational fisheries management processes may help foster positive relationships between anglers and scientists and may provide forums for ongoing communication, not necessarily solely about recreational fishing statistics. Wilson (1999) suggests four types of approaches for angler–scientist interactions, based on a review of North American cases. Two of these are particularly pertinent to recreational fisheries statistics. The first of these types of fisherman–scientist partnerships is the "deference model" in which anglers collect data for use by scientists, but the analysis and interpretation is left to the scientists. Examples of these types include angler catch and participation diaries (e.g., Connelly and Brown, 1995) and various tagging studies (e.g., Kohler et al., 1998). Volunteer angler logbooks (e.g., for striped bass; see Appendix C) also help foster this connection between anglers and scientists and provide a useful database that can be used for estimation of certain fishery statistics, such as discards. In the commercial sector, this approach has also included at-sea collaborations in which scientists are aboard commercial fishing vessels from which they collect data.

The second type of collaboration suggested by Wilson (1999) is that of traditional ecological knowledge, a concept from the study of the value of indigenous knowledge in informing natural resource management decisions. The focus here is to acknowledge that local anglers possess a type of knowledge that is different from science-based knowledge, but that knowledge is valid and useful, based on long-term observation and experience (Neis et al., 1996; Pederson and Hall-Arber, 1999). Traditional ecological knowledge has been used in recreational as well as commercial fishery systems (Sutton, 1999).

Wilson (1999) notes that perhaps the most established cooperative research group is the Fishermen and Scientist Research Society in Nova Scotia, which was initiated in 1993. Building trust between fishermen and scientists was noted as both a crucial challenge to overcome and an important accomplishment achieved through the partnership (King, 1999). Although the Fishermen and Scientist Research Society focuses on commercial fishermen, the experience provides lessons for the recreational fishing and scientific communities.

Establish Stakeholder Advisory Groups

Ongoing citizen and angler advisory groups provide a forum to enable managers and scientists to learn about non-scientist, non-manager concerns and perspectives regarding resource management data and decisions, and also provide a learning opportunity for anglers and other stakeholders to become more informed about fishery management issues. For example, Oregon's Department of Fish and Wildlife Sportfishing Advisory Group provides guidance and consultation to fishery managers before they take in-season management actions or otherwise change allowed fishing patterns and regulations for lingcod (see Appendix C). Advisory groups have the potential to contribute to better informed management decisions, as they have for some regional fishery management councils, as well as better relationships between anglers and managers and scientists.

Forge Partnerships with Others to Implement Outreach Activities

The National Sea Grant system within NOAA provides an established infrastructure with an outreach mission. NMFS staff could work more closely to reach stakeholder groups already identified by Sea Grant Extension Educators, and Sea Grant professionals may be enlisted to help facilitate dialogue among data collectors, data analysts and assessment scientists, managers, and the recreational fishing community. Other partnerships may build from Memoranda of Agreement that NMFS has already established for other outreach purposes, such as with the National Marine Educators Association, the American Sportfishing Association, BOAT/U.S., and the International Game Fish Association (National Oceanic and Atmospheric Administration, 2000b).

Insights regarding the development and establishment of outreach programs involving anglers, scientists, and managers can be gained by learning from existing programs. If the marine recreational fishing statistics data-gathering effort moves toward more state-based (or regional) implementation, outreach programs become much more feasible, in that outreach should be aimed "locally" to respond to the needs of angler-group stakeholders on the scales they most often operate (within the state, statewide, or regionally). One example is the Marine Resource Network that "provides a link between the recreational angling community, research, and fisheries managers. Details on research and projects funded with saltwater license revenues are conveyed to the angling community. This network of some 2,000 individuals establishes a system of volunteers to provide support for outreach and education events" (National Oceanic and Atmospheric Administration, 2000b). Another example is the Atlantic Coastal Cooperative Statistical Program that has an advisory committee comprising representatives from the commercial and recreational sectors (National Oceanic and Atmospheric Administration, 2000b).

Target Opinion Leaders and Innovators within Recreational Fishing Communities

Perceptions of the MRFSS (as well as other surveys) are likely driven less by science and available data than by various opinion leaders and "innovators" in the marine fishing community (recreational fishing organization leaders, tournament anglers, well-known anglers, outdoor writers, and tackle manufacturers and other infrastructure providers) who carry particular weight and influence with rank-and-file anglers and whose support therefore is necessary to foster. Neither rank-and-file anglers nor their representatives are likely to embrace better designed surveys on their own without efforts that increase angler knowledge about the surveys and encourage formation of favorable attitudes toward them.

NMFS's efforts to communicate with recreational fishing community constituents should recognize that "innovators" and "early adopters" typically comprise a small proportion of any specific community (about 15 percent, according to Rogers [2003]). Other individuals within the recreational fishing community are likely to be among the "early majority" (34 percent), the "late majority" (34 percent), and the "laggards" (16 percent). Rogers (2003) describes the latter two categories as being "a skeptical group," adopting new ideas reluctantly. Establishing productive relationships with them will take time and effort.

Rogers (2003) identifies five sequential stages in the process of innovation decision making, which apply to outreach with marine recreational and commercial fishing communities. First, individuals need to be exposed to the new and improved surveys and understand how they work. Second, individuals will form either favorable or unfavorable attitudes toward the new surveys and the way in which survey data are used. Third, anglers will need to engage in activities that lead to a decision to accept or reject the new surveys, implying the importance of opportunities to interact regularly in positive, active ways with data collectors and decision makers so that information about how surveys are conducted and used is consistently reinforced. Fourth, implementation occurs when individuals adopt the new survey and the results it provides in support of fisheries management. Finally, the decision to adopt the new innovation (i.e., a new survey approach) must be reinforced by other messages from NMFS that are consistent with what anglers are learning about the operations and value of the new marine recreational fishing survey approach. Anglers may reverse their decision to adopt the new survey and information it provides if they are exposed to messages that conflict with their understandings. Frequent, consistent, positive, and interactive messages are required to promote continued adoption.

Contact Anglers Directly Onsite, Electronically, and Through Mass Media

Much of what has been said already emphasizes the importance of building relationships and of consistent, positive interactions over time. NMFS staff should not expect that anglers will come to them. They should be prepared to go to the anglers. Reaching opinion leaders, as well as rank-and-file, can be done by regular attendance at angler association meetings, even to the point of becoming a regular agenda item at meetings. The more frequent, more consistent, and more collegial an outreach message is, the more positively it will be received by the stakeholder groups of interest. As noted by the National Oceanic and Atmospheric Administration (2000b), anglers want "timely feedback to the recreational fishing community in terms of survey findings. They are often asked to provide the data, but they do not hear anything about results. From their point of view, it seems as if the data go into a dark hole. They want systems set up where the information is reported back to them." Regular dialogue with angler groups addresses this need. Working with species-specific constituent organizations may be a feasible approach to address specific questions that link data collection efforts with concerns about stock management.

An alternative to meeting anglers onsite is to communicate with anglers directly through electronic means and through mass media. The availability of a list frame for sampling purposes (e.g., state saltwater fishing licenses, a national registry) would also provide an opportunity for fisheries managers and scientists to identify, and thus contact, individual anglers for purposes other than data gathering. Fisheries managers and scientists could regularly produce articles for mainstream, highly subscribed recreational fishing magazines and other outlets.

Provide Access to Data and Training on How to Use Them

The National Research Council (2000) recommends that NMFS use its internet capability, in a more interactive sense, to provide easily accessible and understandable data visualizations (e.g., graphic plots, maps, pictures), as well as providing the ability to access and manipulate data on marine fisheries. Since 2000, the technology to provide easy access to user-specified data requests has improved, and data users have come to expect convenient access to data summaries. Federal statistical agencies have put considerable resources into developing user-friendly interfaces on their websites. Examples of these include the U.S. Census Bureau's fact finder (U.S. Census Bureau, 2006) and the National Center for Education Statistics' National Assessment of Educational Progress Data Explorer (National Center for Education Statistics, 2006). Developing such systems should be done in conjunction with stakeholder groups who are likely to use them to ensure that websites are user-friendly and to explain appropriately the limitations and purposes of the data presented. Other recommendations (National Research Council, 2000) include conducting fishery assessment and management simulations with real data within a workshop, and including a variety of stakeholder groups who could hear each others' interpretations, concerns, and perspectives (e.g., anglers, commercial fishermen, environmental advocates).

CONCLUSIONS AND RECOMMENDATIONS

The success of a communication and outreach program depends on identifying the needs of various stakeholder groups in the marine recreational fishing community and then responding to those needs. Thus far, there has been a lack of confidence with the MRFSS. Improved communication and outreach on data collection efforts, interpretation, and use will improve the credibility of the MRFSS. **It is difficult for individual anglers to see the effects of recreational fishing on their target species and to distinguish daily and seasonal fluctuations from trends.** As a result, no matter how well designed and implemented a marine recreational survey is, it will not succeed fully without the cooperation of anglers. **Unless anglers believe that the survey is well designed and implemented and that it is being used intelligently to address appropriate management issues, they are unlikely to participate. In particular, anglers need to have a basic understanding of the relationship between a statistically based sampling scheme and the frequency with which each of them is (or is not) contacted by a data collector. If anglers believe that their input is influencing the design and use of surveys, they are more likely to be satisfied with those surveys than otherwise. If anglers understand the basic purposes of recreational fishing survey data, the decisions to which these data are being applied, and how those data are interpreted and used, they are more likely to feel confident that the approaches used are legitimate and are more likely to participate willingly and provide valid information. The MRFSS managers**

should advise anglers and data users on the constraints that apply to the use of the data for various purposes. Managers and anglers also should be informed clearly about any limitations of the data. The MRFSS process and mandate is also in need of a formalized method for periodic evaluation to ensure that the program continually evolves to meet the needs of the stakeholder groups.

In addition, the committee feels that a meaningful dialogue between managers and anglers will require more interactions and better relationships between the two groups. **Outreach and communication should be institutionalized as part of an ongoing MRFSS program so their importance is acknowledged and appropriate expertise can be developed. Angler associations should be engaged as partners with survey managers through workshops, data collection, survey design, and participation in survey advisory groups.** Partnerships with other programs, particularly those with existing outreach programs (such as Sea Grant), could facilitate outreach efforts. Many National Research Council and other reports stress the importance of making use of local and traditional knowledge and capacity building and the involvement of local communities in knowledge-gathering and dissemination activities. Those recommendations apply, as well, to the recreational fishing community and can be used as a resource for future MRFSS outreach efforts.

References

Äas, O., W. Haider, and L. Hunt. 2000. Angler responses to harvest regulations in Engerdal, Norway: A conjoint based choice modeling approach. *North American Journal of Fisheries Management* 20:940-950.

American Statistical Association. 2006a. *ASA Committees*. [Online]. Available: http://www.amstat.org/comm/index.cfm?fuseaction=main [2006, February 9].

American Statistical Association. 2006b. *Research Fellowship Programs*. [Online]. Available: http://www.amstat.org/research_grants/index.cfm?fuseaction=main [2006, February 9].

Atlantic States Marine Fisheries Commission. 2003. *Fishery Management Report No. 41 of the Atlantic States Marine Fisheries Commission: Amendment 6 to the Interstate Fishery Management Plan for Atlantic Striped Bass February 2003*. Atlantic States Marine Fisheries Commission, Washington, DC.

Atlantic States Marine Fisheries Commission. 2004. *2004 Stock Assessment Report for Atlantic Striped Bass: Catch-at-Age Based VPA and Tag Release/Recovery Based Survival Estimation*. Prepared by the Striped Bass Technical Committee for the Atlantic Striped Bass Management Board, Atlantic States Marine Fisheries Commission, Washington, DC.

Ault, J.S., S.G. Smith, D.A. Guillermo, and E. Franklin. 2003. *Florida Hogfish Fishery Stock Assessment*. [Online]. Available: http://bonefish.rsmas.miami.edu/femar/Publications/hogfish_stock_assessment_2003.pdf [2006, February 7].

Battese, G.E., R.M. Harter, and W.A. Fuller. 1988. An error-components model for prediction of county crop areas using survey and satellite data. *Journal of the American Statistical Association* 83:28-36.

Bohnsack B.L., R.B. Ditton, J.R. Stoll, R.J. Chen, R. Novak, and L.S. Smutko. 2002. The economic impacts of the recreational bluefin tuna fishery in Hatteras, North Carolina. *North American Journal of Fisheries Management* 22:165-176.

Breidt, F.J. 2002. National Resources Inventory (NRI), US. In: *Encyclopedia of Environmetrics* (vol. 3), El-Shaarawi, A.H. and W.W. Piegorsch (eds.). John Wiley and Sons, Inc., Hoboken, NJ.

Breidt, F.J. and W.A. Fuller. 1999. Design of supplemented panel surveys with application to the National Resources Inventory. *Journal of Agricultural, Biological, and Environmental Statistics* 4:391-403.

Breidt, F.J. and J.D. Opsomer. 2000. Local polynomial regression estimators in survey sampling. *Annals of Statistics* 28:1026-1053.

Brockwell, P.J. and R.A. Davis. 1991. *Time Series: Theory and Methods* (2nd ed.). Springer-Verlag, New York, NY.

Cameron, T.A. and D.D. Huppert. 1989. OLS versus ML estimation of non-market resource values with payment card interval data. *Journal of Environmental Economics and Management* 17(3):230-246.

Cantwell, P.J. 1990. Variance formulas for composite estimation in rotation designs. *Survey Methodology* 16:153-163.

Chhikara, R.S. and L.Y. Deng. 1992. Estimation using multiyear rotation design sampling in agricultural surveys. *Journal of the American Statistical Association* 87:924-932.

Cochran, W.G. 1977. *Sampling Techniques* (3rd ed.). John Wiley and Sons, Inc., Hoboken, NJ.

Coleman, F.C., W.F. Figueira, J.S. Ueland, and L.B. Crowder. 2004. The impact of United States recreational fisheries on marine fish populations. *Science* 305:1958-1959.

Colorado State University. 2005. *STARMAP Program*. [Online]. Available: http://www.stat.colostate.edu/%7Ensu/starmap/program.html [2006, February 9].

Connelly, N.A. and T.B. Brown. 1995. Use of angler diaries to examine biases associated with 12-month recall on mail questionnaires. *Transactions of the American Fisheries Society* 124:413-442.

Cooke, S.J. and I.G. Cowx. 2004. The role of recreational fishing in global fish crises. *Bioscience* 54(9):857-859.

Coughenower, D. and C. Blood. 1997. *Flat Out Facts about Halibut*. Alaska Sea Grant College Program, University of Alaska, Fairbanks.

Cressie, N.A.C. 1993. *Statistics for Spatial Data* (revised ed.). John Wiley and Sons, Inc., Hoboken, NJ.

Dauk, P.C. and C.J. Schwarz. 2001. Catch estimation using complex surveys with no randomization in the effort survey. *Biometrics* 57:461-468.

Dietzenbacher, E. and M.L. Lahr (eds.). 2004. *Wassily Leontief and Input-Output Economics*. Cambridge University Press, New York, NY.

Diewert, R.E., D.A. Nagtegaal, and K. Hein. 2005. A comparison of the results of the 1998 Georgia Strait creel survey with an independent observer program. *Canadian Manuscript Report of Fisheries and Aquatic Sciences No. 2716*. University of New Brunswick.

Dillman, D.A. 2000. *Mail and Internet Surveys: The Tailored Design Method* (2nd ed.). John Wiley and Sons, Inc., Hoboken, NJ.

Ditton, R.B., A. Loftus, and J.H. Volstad. 2002. *Review Findings of the ACCSP For-Hire Pilot Survey*. Report to the Atlantic Coastal Cooperative Statistics Program, Washington, DC.

Diodati, P.J. and R.A. Richards. 1996. Mortality of striped bass hooked and released in salt water. *Transactions of the American Fisheries Society* 125:300-307.

Essig, R.J. and M.C. Holliday. 1991. Development of a recreational fishing survey: The marine recreational fishery statistics survey case study. *American Fisheries Society Symposium* 12:162-176.

Federal Register. 2002. Fisheries of the Caribbean, Gulf of Mexico, and south Atlantic; coastal migratory pelagic resources of the Gulf of Mexico and south Atlantic; reef fish resources of the Gulf of Mexico; charter vessel and headboat permit moratorium. *Federal Register* 67(39):8926-8927.

Florida Fish and Wildlife Conservation Commission. 2005. *Recreational Saltwater Fishing License(s) Requirements*. [Online]. Available: http://www.marinefisheries.org/license.htm#rec [2006, March 21].

Foster, D.G. 2004. *1999-2003 North-Central and Western Gulf of Mexico BRD Performance*. Report to the Gulf of Mexico Fishery Management Council. National Oceanic and Atmospheric Administration, Southeast Fisheries Science Center, Pascagoula, MS.

Gautam, A. and S.R. Steinback. 1998. Valuation of recreational fisheries in the northeast U.S. striped bass: A case study. In *Recreational Fisheries: Social, Economic, and Management Aspects*, Hickley, P. and H. Tompkins (eds.). Fishing News Books, Oxford, England.

Ghosh, M. and J.N.K. Rao. 1994. Small area estimation: An appraisal. *Statistical Science* 9:55-93.

Goodyear, C.P. 1997. Fish age determination from length: An evaluation of three methods using simulated data. *Fisheries Bulletin* 95:39-46.

Green, L.M., H.R. Osburn, G.C. Matlock, and R.B. Ditton. 1991. Development of a social and economic questionnaire for on-site interviews of marine recreational anglers in Texas. *American Fisheries Society Symposium* 12:406-412.

Groves, R.M., F.J. Fowler, M.P. Couper, J.M. Lepkowski, E. Singer, and R. Tourangeau. 2004. *Survey Methodology*. John Wiley and Sons, Inc., Hoboken, NJ.

Gulf of Mexico Fishery Management Council. 2003. *Draft Environmental Impact Statement for the Generic Essential Fish Habitat Amendment to the Following Fishery Management Plans of the Gulf of Mexico: Shrimp Fishery of the Gulf of Mexico, Red Drum Fishery of the Gulf of Mexico, Reef Fish Fishery of the Gulf of Mexico, Stone Crab Fishery of the Gulf of Mexico, Coral and Coral Reef Fishery of the Gulf of Mexico, Spiny Lobster Fishery of the Gulf of Mexico and South Atlantic, and Coastal Migratory Pelagic Resources of the Gulf of Mexico and South Atlantic*. Gulf of Mexico Fishery Management Council, Tampa, FL.

Gulf of Mexico Fishery Management Council. 2004a. *Final Amendment 22 to the Reef Fish Fishery Management Plan to Set Red Snapper Sustainable Fisheries Act Targets and Thresholds, Set a Rebuilding Plan, and Establish Bycatch Reporting Methodologies for the Reef Fish Fishery*. [Online].

Available: http://www.gulfcouncil.org/downloads/Amend%2022%20Final %2070204.pdf [2006, February 7].

Gulf of Mexico Fishery Management Council. 2004b. *SEDAR 7: Gulf of Mexico Red Snapper—Red Snapper Data Workshop Report*. [Online]. Available: http://www.gulfcouncil.org/downloads/SEDAR%207%20DW_FINALREP ORT_RS.pdf [2006, February 7].

Gulf of Mexico Fishery Management Council. 2005a. *SEDAR 7: Southeast Data, Assessment, and Review—Stock Assessment Report of SEDAR 7 Gulf of Mexico Red Snapper*. [Online]. Available: http://www.gulfcouncil.org/ downloads/SEDAR7%20AW%20report%20main.pdf [2006, February 7].

Gulf of Mexico Fishery Management Council. 2005b. *Gulf of Mexico Fishery Management Council Report to the National Research Council on the Recreational Fishing Survey Methods July 2005*. [Online]. Available: http://www.gulfcouncil.org/downloads/NRC-Recreational-Letter.pdf [2006, February 7].

Gulf of Mexico Fishery Management Council. 2005c. *Public Hearing Draft Amendment to the FMPs for Reef Fish (Amendment 25) and Coastal Migratory Pelagics (Amendment 17) for Extending the Charter Vessel/Head Boat Moratorium, Including SEIS/RIR/IRFA*. Gulf of Mexico Fishery Management Council, Tampa, FL.

Gulland, J.A. 1983. *Fish Stock Assessment: A Manual of Basic Methods* (vol. 1). Series on Food and Agriculture. Food and Agriculture Organization/John Wiley and Sons, Inc.

Guthrie, D., J.M. Hoenig, M. Holliday, C.M. Jones, M.J. Mills, S.A. Moberley, K.H. Pollock, and D.R. Talhelm (eds.). 1991. Creel and angler surveys in fisheries management. *American Fisheries Society Symposium 12*. American Fisheries Society.

Haab, T.C., J.C. Whitehead, and K.E. McConnell. 2001. *The Economic Value of Marine Recreational Fishing in the Southeast United States*. National Oceanic and Atmospheric Administration.

Hardie, D.C., D.A. Naagtegaal, and L. Nagy. 1998. Strait of Georgia sport fishery creel survey statistics for salmon and ground fish. *Canadian Fisheries and Aquatic Sciences Report*. University of New Brunswick.

Hartley, H.O. 1962. Multiple frame surveys. *Proceedings of the Social Statistics Section of the American Statistical Association*. American Statistical Association, Washington, DC.

Henry, G.W. 2002. The national recreational and indigenous fishing survey. *Proceedings of the Third World Recreational Fishing Conference*. May 21-24, 2002, Darwin, Australia.

Hicks, R.L. 2002. *Stated Preference Methods for Environmental Management: Recreational Summer Flounder Angling in the Northeastern United States*. National Oceanic and Atmospheric Administration, Silver Spring, MD.

Holiman, S.G. 1996. *Estimating Recreational Effort Using the Marine Recreational Fisheries Statistics Survey Data*. National Oceanic and Atmospheric Administration Technical Memorandum.

Holland, S.M. and R.B. Ditton. 1992. Fishing trip satisfaction: A typology of anglers. *North American Journal of Fisheries Management* 12:28-33.

Huppert, D.D. 1989. Measuring the value of fish to anglers: Application to central California anadromous species. *Marine Resource Economics* 6(2):89-108.

Jagielo, T.H., F.R. Wallace, and Y.W. Cheng. 2003. *Assessment of Lingcod (Opiodon elongates) for the Pacific Fishery Management Council in 2003.* [Online]. Available: http://www.pcouncil.org/groundfish/gfsafe0604/lingcod.pdf [2006, February 7].

Jentoft, S., B.J. McCay, and D.C. Wilson. 1998. Social theory and fisheries co-management. *Marine Policy* 22:423-436.

Johnston, R.S. 1992. *Fisheries Development, Fisheries Management, and Externalities*. World Bank Discussion Paper No. 165. The World Bank, Washington, DC.

King, J.R. and D.R. Haggarty. 2004. An examination of recapture rates of lingcod as a potential source of bias in recreational catch per unit effort (CPUE) indices. *Canadian Manuscript Report of Fisheries and Aquatic Sciences No. 2670*. University of New Brunswick.

King, P. 1999. Practitioners' profile: The fishermen and scientists research society. *Common Property Resource Digest* 49:9-11.

Kohler, N.E., J.G. Casey, and P.A. Turner. 1998. NMFS Cooperative Shark Tagging Program, 1962-1993: An atlas of shark tag and recapture data. *Marine Fisheries Review* 60(2):1-87.

Layton, D.F., G.M. Brown, Jr., and M.L. Plummer. 1999. *Valuing Multiple Programs to Improve Fish Populations*. Unpublished document. Department of Ecology, Washington State University, Pullman.

Lee, L.M. 2003. *Population Assessment and Short-Term Stock Projections of the Bluefish (Pomatomus saltatrix)*. Atlantic States Marine Fisheries Commission, Washington, DC.

Legault, C.M. and V.R. Restrepo. 1998. A flexible forward age-structured assessment program. *International Commission for the Conservation of Atlantic Tunas Collective Volume of Scientific Papers* 49:246-253.

Lepkowski, J.M. and M.P. Couper. 2002. Nonresponse in the second wave of longitudinal household surveys. In *Survey Nonresponse*, Groves, R., D. Dillman, J. Eltinge, and R. Little (eds.). John Wiley and Sons, Inc., Hoboken, NJ.

Lockwood, R.N., J. Peck, and J. Oelfke. 2001. Survey of angling in Lake Superior waters at Isle Royale National Park, 1998. *North American Journal of Fisheries Management* 21:471-481.

Loftus, A.J., G.C. Radonski, and R.B. Stone. 1999. *Interactive Outreach with the Atlantic Coast Recreational Fishing Community for the ACCSP*. Final report to the Atlantic Coastal Cooperative Statistics Program, Washington, DC.

Lohr, S.L. 1999. *Sampling Design and Analysis*. Duxbury Press, Pacific Grove, CA.

Louviere, J.J., D. Hensher, and J. Swait. 2000. *Stated Choice Methods: Analysis and Application*. Cambridge University Press, London, England.

Lucy, J. and A.L. Studholme (eds.). 2002. Catch and release in marine recreational fisheries. *American Fisheries Society Symposium 30*. American Fisheries Society.

Lyle, J.M., A.P.M. Coleman, L. West, D. Campbell, and G.W. Henry. 2002. An innovative methodology for the collection of detailed and reliable data in large-scale Australian recreational fishing surveys. In *Recreational Fisheries: Ecological, Economic, and Social Evaluation*, Pitcher, T.J. and C.E. Hollingworth (eds.). Fish and Aquatic Resources Series No 8. Blackwell Science, Oxford, England.

Malvestuto, S.P. 1983. Sampling the recreational fishery. In *Fisheries Techniques*, Nielsen, L.A. and D.L. Johnson (eds.). American Fisheries Society, Bethesda, MD.

McConnell, K.E., I.E. Strand, S. Valdes, and Q.R. Weninger. 1994. *Volume 2: The Economic Value of Mid and South Atlantic Sportfishing*. Department of Agricultural and Resource Economics, Environmental Protection Agency, Washington, DC.

McFadden, D. 2001. Economic choices. *The American Economic Review* 91(3):351-378.

Mitchell, R.C. and R.T. Carson. 1989. *Using Surveys to Value Public Goods: The Contingent Valuation Method*. Resources for the Future, Washington, DC.

Munger, L., G. Shepard, and W. Laney. 2005. *2005 Review of the Atlantic States Marine Fisheries Commission's Fishery Management Plan for Atlantic Striped Bass (Morone saxatilis)*. Prepared by the Striped Bass Plan Review Team, Atlantic States Marine Fisheries Commission, Washington, DC.

National Center for Education Statistics. 2006. *NAEP Data Explorer*. [Online]. Available: http://nces.ed.gov/nationsreportcard/nde/ [2006, February 21].

National Oceanic and Atmospheric Administration. 2000a. *Other Surveys*. [Online]. Available: http://www.st.nmfs.gov/st1/econ/survey_o.html [2006, April 24].

National Oceanic and Atmospheric Administration. 2000b. *RecFish 2000: Managing Marine Recreational Fisheries in the 21st Century—Meeting the Needs of Managers, Anglers, and Industry*. [Online]. Available: http://www.nmfs.noaa.gov/irf/RecFish2000.pdf [2006, February 9].

National Oceanic and Atmospheric Administration. 2005a. *Fisheries of the United States: 2004*. National Oceanic and Atmospheric Administration, Silver Spring, MD.

National Oceanic and Atmospheric Administration. 2005b. *NOAA Recreational Fisheries Statistics Program*. [Online]. Available: http://www.st.nmfs.gov/st1/recreational/index.html [2006, February 7].

National Research Council. 1999. *Sustaining Marine Fisheries*. National Academy Press, Washington, DC.

National Research Council. 2000. *Improving the Collection, Management, and Use of Marine Fisheries Data.* National Academy Press, Washington, DC.

National Research Council. 2002. *Science and Its Role in the National Marine Fisheries Service.* National Academy Press, Washington, DC.

National Research Council. 2004. *Improving the Use of the "Best Scientific Information Available" Standard in Fisheries Management.* National Academies Press, Washington, DC.

National Science Foundation. 2005. *Methodology, Measurement, and Statistics (MMS).* [Online]. Available: http://nsf.gov/pubs/2004/nsf04504/nsf04504.pdf [2006, February 9].

Neis, B., L. Felt, D.C. Schneider, R. Haedrich, J. Hutchings, and J. Fischer. 1996. *Northern Cod Stock Assessments: What Can Be Learned From Interviewing Resource Users?* DFO Atlantic Fisheries Research Document 96/45, Fisheries and Oceans, Canada.

Nieuwenbroek, N.J. 1991. Precision of net change in a rotating panel survey. *The Statistician* 40:195-201.

Northeast Fisheries Science Center. 2003. *Report of the 36th Northeast Regional Stock Assessment Workshop (36th SAW): Stock Assessment Review Committee (SARC) Consensus Summary of Assessments.* Northeast Fisheries Science Center Reference Document 03-06, National Marine Fisheries Service, Woods Hole, MA.

Nusser, S.M. and J.J. Goebel. 1997. The National Resources Inventory: A long-term multi-resource monitoring program. *Environmental and Ecological Statistics* 4:181-204.

Oh, C.O., R.B. Ditton, B. Genter, and R. Riechers. 2005. A stated preference choice approach to understanding angler preferences for management options. *Human Dimensions of Wildlife* 10(3):173-186.

Oregon State University. 2005. *DAMARS.* [Online]. Available: http://oregonstate.edu/dept/statistics/epa_program/index.html [2006, February 9].

Pacific Fishery Management Council. 2003. *Lingcod: STAR Panel Meeting Report.* [Online]. Available: http://www.pcouncil.org/groundfish/gfsafe0604/lingcodstar.pdf [2006, March 24].

Pacific Fishery Management Council. 2004. *Pacific Coast Groundfish Fishery Management Plan for the California, Oregon, and Washington Groundfish Fishery as Amended through Amendment 17.* [Online]. Available: http://www.pcouncil.org/groundfish/gffmp/fmpthru17.pdf [2006, February 7].

Pacific Fishery Management Council. 2006. *2005 Stock Assessments and STAR Panel Reports.* [Online]. Available: http://www.pcouncil.org/groundfish/05gfstokasmts.html [2006, February 7].

Parrack, N.C. 1986a. *A Review of Gulf of Mexico Red Snapper Age and Growth.* National Marine Fisheries Service.

Parrack, N.C. 1986b. *Review and Update of Gulf of Mexico Red Snapper Biometrics.* National Marine Fisheries Service.

Parrack, N.C. and D.B. McClellan. 1986. *Trends in Gulf of Mexico Red Snapper Population Dynamics, 1979–85*. National Marine Fisheries Service.

Parsons, G.R. 2003. The travel cost model. In *A Primer on Nonmarket Valuation*, Champ, P.A., K.J. Boyle, and T.C. Brown (eds.). Kluwer Academic Publishers, Boston, MA.

Pauly, D. 1998. Rationale for reconstructing catch time series. *European Commission Fisheries Cooperation Bulletin* 11(2):4-7.

Pederson, J. and M. Hall-Arber. 1999. Fish habitat: A focus on New England fishermen's perspectives. *American Fisheries Society Symposium* 22:188-211.

Policansky, D. 2002. Catch-and-release recreational fishing: A historical perspective. In *Recreational Fisheries: Ecological, Economic, and Social Evaluation*, Pitcher, T.J. and C. Hollingworth (eds.). Blackwell Science, Oxford.

Pollock, K.H. 2002. Recreational angler surveys: The interaction of scale and optimal contact methods for effort and catch estimation. *Proceedings of the Third World Recreational Fishing Conference*. May 21-24, 2002, Darwin, Australia.

Pollock, K.H., C.M. Jones, and T.L. Brown. 1994. *Angler Survey Methods and their Applications in Fisheries Management*. American Fisheries Society Special Publication 25, Bethesda, MD.

Porch, C.E. 2004a. *An Alternative Assessment of the Red Snapper (Lutjanus campechanus) Fishery in the U.S. Gulf of Mexico Using a Spatially-Explicit Age-Structured Assessment Model: Preliminary Results*. Southeast Fisheries Science Center, Sustainable Fisheries Division, Miami, FL.

Porch, C.E. 2004b. *An Age-Structured Assessment Model for Red Snapper that Allows for Multiple Stocks, Fleets, and Habitats*. Southeast Fisheries Science Center, Sustainable Fisheries Division, Miami, FL.

Ralston, S. and E.J. Dick. 2003. *The Status of Black Rockfish (Sebastes melanops) off Oregon and Northern California in 2003*. [Online]. Available: http://www.pcouncil.org/groundfish/gfsafe0803/black.pdf [2006, February 7].

Rao, J.N.K. 2003. *Small Area Estimation*. John Wiley and Sons, Inc., Hoboken, NJ.

Richards, L.J., J.T. Schnute, and C.M. Hand. 1990. A multivariate maturity model with a comparative analysis of three lingcod (*Ophiodon elongatus*) stocks. *Canadian Journal of Fisheries and Aquatic Sciences* 47:948-959.

Robins, C.R., G.C. Rey, and J. Douglass. 1986. *A Field Guide to Atlantic Coast Fishes*. Houghton Mifflin Co., New York, NY.

Robson, D.S. 1960. An unbiased sampling and estimation procedures for creel censuses of fishermen. *Biometrics* 16:261-277.

Robson, D.S. 1961. On the statistical theory of a roving creel census. *Biometrics* 17:415-437.

Robson, D.S. and C.M. Jones. 1989. The theoretical basis of an access site angler survey design. *Biometrics* 45:83-98.

Rogers, E.M. 2003. *The Diffusion of Innovations* (5th ed.). The Free Press, New York, NY.

Särndal, C.E., B. Swensson, and J. Wretman. 1992. *Model Assisted Survey Sampling*. Springer-Verlag, New York, NY.

Schirripa, M.J. 1999. Management trade-offs between the directed and undirected fisheries of red snapper (*Lutjanus campechanus*) in the U.S. Gulf of Mexico. *Proceedings of the International Conference on Integrated Fisheries Monitoring*. Sydney, Australia, 1-5 February 1999. Food and Agriculture Organization, Rome, Italy.

Schirripa, M.J. and C.M. Legault. 1999. *Status of the Red Snapper in U.S. Waters of the Gulf of Mexico: Updated through 1998*. [Online]. Available: http://www.gulfcouncil.org/downloads/RSAssess99.pdf [2006, February 7].

Southeast Fisheries Science Center. 2002. *Status of Red Grouper in United States Waters of the Gulf of Mexico During 1986–2001*. [Online]. Available; http://www.gulfcouncil.org/Beta/GMFMCWeb/downloads/Gulf-RedGrouper-Assessment-1986to2001-revised12132002.pdf [2006, May 1].

Steinback, S.R. 1999. Regional economic impact assessments of recreational fisheries: An application of the IMPLAN modeling system to marine party and charter boat fishing in Maine. *North American Journal of Fisheries Management* 19:724-736.

Stephens, A. and A. MacCall. 2004. A multispecies approach to subsetting logbook data for purposes of estimating CPUE. *Fisheries Research* 70:299-310.

Sutton, T. 1999. *Local Knowledge of a Unique Population of Atlantic Salmon: Implications for Community-Based Management of Recreational Fisheries in Newfoundland and Labrador*. Department of Wildlife and Fisheries Science, Texas A&M University, College Station, TX.

Tiller, R. 1992. Time series modeling of sample survey data from the U.S. Current Population Survey. *Journal of Official Statistics* 8:149-66.

University of Washington. 2004. *Coleraine: A Generalized Age Structured Stock Assessment Model*. [Online]. Available: http://fish.washington.edu/research/coleraine/ [2006, May 26].

U.S. Census Bureau. 2001. *Current Population Survey*. [Online]. Available: http://www.bls.census.gov/cps/cpsmain.htm [2006, February 7].

U.S. Census Bureau. 2006. *U.S. Census Bureau: Fact Finder*. [Online]. Available: http://factfinder.census.gov/home/saff/main.html?_lang=en [2006, February 21].

U.S. Department of Agriculture. 1999. *Summary Report: 1997 National Resources Inventory*. Natural Resources Conservation Service, Washington, DC, and Statistical Laboratory, Iowa State University, Ames.

U.S. Department of Justice. 2005. *Bureau of Justice Statistics Criminal Victimization Data Collections*. [Online]. Available: http://www.ojp.usdoj.gov/bjs/cvict.htm#ncvs [2006, February 7].

Volstad, J.H., K.H. Pollock, and W. Rickus. In press. Some new methodological developments in the design and analysis of aerial-access angler surveys:

The Delaware River angler survey 2002. *Transactions of the American Fisheries Society*.

Watson, R., S. Guénette, P. Fanning, and T.J. Pitcher. 2000. The basis for change: Part 1—Reconstructing fisheries catch and effort data. In *Methods for Evaluating the Impacts of Fisheries on North Atlantic Ecosystems*, Pauly, D. and T.J. Pitcher (eds.). Fisheries Centre Research Reports, University of British Columbia, Vancouver, Canada.

Wilson, C.A. and D.L. Nieland. 2001. Age and growth of red snapper, *Lutjanus campechanus*, from the northern Gulf of Mexico off Louisiana. *Fisheries Bulletin* 99:653-664.

Wilson, D.C. 1999. *Fisheries Science Collaborations: The Critical Role of the Community*. Research Publication No. 45. Institute for Fisheries Management and Coastal Community Development, North Sea Centre, Hirtshals, Denmark.

Wilson, D.C., J.R. Nielsen, and P. Degnbol (eds.). 2003. *The Fisheries Co-Management Experience: Accomplishments, Challenges, and Prospects*. Kluwer Academic Publishers, Dordrecht, The Netherlands.

Wolter, K. 1979. Composite estimation in finite populations. *Journal of the American Statistical Association* 74:604-613.

Yandle, T. 2003. The challenge of building successful stakeholder organizations: New Zealand's experience in developing a fisheries co-management regime. *Marine Policy* 27:179-192.

Appendix A

Committee and Staff Biographies

COMMITTEE

Patrick J. Sullivan (*Chair*) is an associate professor in the Department of Natural Resources at Cornell University. Dr. Sullivan earned a Ph.D. in biostatistics in 1988 from the University of Washington. His research focuses on the assessment and management of fisheries resources and the statistical modeling of biological systems.

F. Jay Breidt is a professor of statistics at Colorado State University. Dr. Breidt earned a Ph.D. in statistics in 1991 from Colorado State University. His research focuses on non-Gaussian linear time series models, environmental monitoring, and nonparametric regression in surveys.

Robert B. Ditton is a professor of wildlife and fisheries sciences at Texas A&M University. Dr. Ditton earned a Ph.D. in recreation and park administration in 1969 from the University of Illinois. His research focuses on the sociology of natural resources with special attention to the human dimensions of fisheries. He is a previous member of the Ocean Studies Board.

Barbara A. Knuth is the Chair of the Department of Natural Resources at Cornell University. Dr. Knuth earned a Ph.D. in fisheries and wildlife sciences in 1986 from Virginia Polytechnic Institute and State University. Her research focuses on risk perception, communication, and management of fisheries affected by chemical contaminants.

Bruce M. Leaman is the Executive Director of the International Pacific Halibut Commission and an affiliate professor in the School of Aquatic and Fishery Sciences at the University of Washington. Dr. Leaman earned a Ph.D. from the University of British Columbia. His research focuses on reproductive and evolutionary biology of long-lived fishes; stock assessment; and fisheries governance, management, and harvest policy.

Victoria M. O'Connell is the Groundfish Project Leader for the Southeast Region of the Alaska Department of Fish and Game. Ms. O'Connell earned a B.S. in fisheries from the University of Washington. Her research focuses on the management of southeast Alaskan commercial groundfish fisheries, including longline and pot fisheries for sablefish, longline fisheries for rockfish, and jig and troll fisheries for lingcod.

George R. Parsons is a professor in the Graduate College of Marine Studies and Department of Economics at the University of Delaware. Dr. Parsons earned a Ph.D. in economics in 1985 from University of Wisconsin-Madison. His research focuses on the application of random utility models to recreational fishing and hunting in Delaware; beach use on the Mid-Atlantic, Texas, and New England coasts; and diving in the Caribbean.

Kenneth H. Pollock is a professor of zoology, biomathematics, and statistics at North Carolina State University. Dr. Pollock earned a Ph.D. in biometry from Cornell University. His research focuses on sampling methods used in wildlife and fisheries science and management, including the design and analysis of recreational angler surveys.

Stephen J. Smith is a research scientist with the Bedford Institute of Oceanography. Mr. Smith earned his B.Sc. degree in marine biology and M.Sc. in statistics from the University of Guelph. His research focuses on the application and design of surveys for marine populations, population dynamics models, and spatial analysis of species distributions in conjunction with data on geology and oceanographic conditions.

S. Lynne Stokes is a professor of statistical science at Southern Methodist University. Dr. Stokes earned a Ph.D. in statistics in 1976 from the University of North Carolina. Her research focuses on sampling methods, modeling of non-sampling errors in surveys, and disclosure limitation methods.

STAFF

Christine Blackburn was a program officer with the Ocean Studies Board until early 2006. She earned a Ph.D. in oceanography from the Scripps Institution of Oceanography. Since receiving her doctorate, Dr. Blackburn has been working in science policy. In 2003, she became a policy associate for the U.S. Commission on Ocean Policy, where she helped prepare the Commission's report. Dr. Blackburn is currently an ocean program analyst with the California Coastal Conservancy.

David Policansky is a scholar and the director of the Program in Applied Ecology and Natural Resources in the Board on Environmental Studies and Toxicology. Dr. Policansky earned a Ph.D. in biology from the University of Oregon. His areas of expertise include genetics; evolution; and ecology, including the effects of fishing on fish populations, ecological risk assessment, natural resource management, and how science is used in informing policy.

Susan Park is an associate program officer with the Ocean Studies Board. She earned a Ph.D. in oceanography in 2004 from the University of Delaware. Her dissertation focused on the range expansion of the nonnative Asian shore crab *Hemigrapsus sanguineus*. In the summer of 2002, she participated in the Christine Mirzayan Science and Technology Graduate Policy Fellowship with the Ocean Studies Board. Prior to joining the Ocean Studies Board in 2006, Dr. Park worked on aquatic invasive species management with the Massachusetts Office of Coastal Zone Management and the Northeast Aquatic Nuisance Species Panel.

Jodi Bostrom is a research associate with the Ocean Studies Board. She earned a B.S. in zoology in 1998 from the University of Wisconsin-Madison. Since starting with the Ocean Studies Board in May 1999, Ms. Bostrom has worked on several studies pertaining to fisheries, marine mammals, nutrient over-enrichment, and ocean exploration. She will earn an M.S. in environmental science from American University in December 2006.

Phillip Long was a program assistant for the Ocean Studies Board and recently became a senior program assistant with the Board on Physics and Astronomy. He earned a B.S. in chemistry and a B.A. in history from the University of Portland. Before coming to the National Academies, Mr. Long worked as a medical research assistant at the Oregon Health Sciences University.

Carrie Wall is a graduate research assistant with the Institute for Marine Remote Sensing at the University of South Florida. She earned an M.S. in biological oceanography in 2006 from the University of South Florida. Ms. Wall has worked on research projects pertaining to marine mammals, environmental education, wetlands restoration, sea turtles, fisheries, ocean policy, and remote sensing.

Appendix B

Existing Marine Recreational Fisheries Surveys

Currently, there are several different surveys of marine recreational fishing conducted throughout the United States (Table B.1). There are supplemental surveys that were created to better analyze a specific sector (e.g., For-Hire Survey [FHS]) and others that better sample specific types of fishing (e.g., Large Pelagic Survey [LPS]). A quick overview of the different surveys is provided below.

MARINE RECREATIONAL FISHERIES STATISTICS SURVEYS

The Marine Recreational Fisheries Statistics Survey (MRFSS) was implemented by the National Marine Fisheries Service in 1979 and was conducted for all recreational fisheries along the Atlantic and Pacific coast, in the Gulf of Mexico and Caribbean Sea, and off Hawaii. The MRFSS is separated into two-month periods called "waves." January and February are wave 1, March and April are wave 2, and so on.

The MRFSS is designed to determine annual, regional harvest estimates and to provide continuous, coastwide monitoring of fish stocks. Total angler fishing trips and total angler catches by species, including removals and catch released alive, are recorded for the annual, regional estimates. Fishing effort and catch per unit effort (CPUE) are recorded for coastwide monitoring. Fishing effort is determined from coastal household telephone surveys, which collect data for each household by recording the number of residents who fished in the last two months; for each angler by recording the number of fishing trips (days) in the last

TABLE B.1 Overview of the Marine Recreational Fishing Surveys that Operate in the United States

Survey Name	Geographic Area	Effort Survey Method(s)	Catch per Unit Effort Survey Method(s)	Catch Survey Method(s)	Mode(s) of Fishing	Fish Species
Alaska Sport Fish Statewide Harvest Survey	Alaska Southeast Alaska	Household directory mail survey (angler license)	Household directory mail survey (angler license) catch per household	Household directory mail survey (angler license) Creel access survey (boat anglers)	All	All species
California Recreational Fisheries Survey	California	Onsite roving instantaneous boat counts	Port-based boat-trip intercept survey		Private boats	All species
			Access-point intercept survey		Pier and dock fishing	
		Angler directory telephone survey (angler license)	Access-point intercept survey		Pier and dock fishing and private and rental boats	All species
			Access-point (roving-creel) intercept survey		Beach and bank fishing	All species

Catch Card Survey	Washington			Mandatory catch card reports	All boats and shore fishing	Salmon and halibut
	North Carolina and Maryland			Mandatory catch card reports	All boats	Bluefin tuna and marlin
For-Hire Survey	Atlantic coast and Gulf of Mexico	Boat directory telephone survey	Access-point intercept survey		Charter and head boats	All species
			At-sea sampling		Head boats	All species
Large Pelagic Survey	Atlantic coast (Virginia to Maine)	Boat directory telephone survey	Access-point intercept survey		Private and charter boats with highly migratory species permits	Large pelagic species
Marine Recreational Fishery Statistics Survey	Atlantic coast and Gulf of Mexico	Random digit dialing household telephone survey	Access-point intercept survey		Shore	All species
					Private and rental boats	All species

Oregon Recreational Boat Survey	Oregon	Onsite boat exit count survey	Port-based boat-trip intercept survey		Ocean-going private boats Ocean-going party and charter boats	All species
Ocean Sampling Program	Washington	Onsite boat entrance count survey	Port-based boat-trip intercept survey		Coastal private boats Coastal party and charter boats	All species
Party Charter Survey (For-Hire Survey)	California	Boat directory telephone survey	Access-point intercept survey At-sea sampling		Party and charter boats	All species
Puget Sound Sampling Program	Washington	Angler directory telephone survey (angler license)	Access-point intercept survey		Puget Sound private boats	All species

Shore and Estuarine Boat Survey	Oregon	Angler directory telephone survey (angler license)	Access-point intercept survey		Shore and inland boats	All species
		Onsite boat exit count survey	Port-based boat-trip intercept survey		Ocean-going private boats	All species
Southeast Head Boat Survey	Gulf of Mexico	Logbook census	Logbook census	Logbook census	Head boats	All species
Texas Marine Recreational Fishing Survey	Gulf of Mexico	Onsite roving instantaneous boat counts	Port-based boat-trip intercept survey		Private and rental boats	
					Charter boats	
Vessel Trip Reports	Atlantic coast	Logbook census	Logbook census	Logbook census	Party, charter, and head boats with federal permits	Federally regulated species

two months; and for each fishing trip by recording state and county of fishing access, private versus public access, mode of fishing, and date and time of return.

CPUE is determined from access-point intercept surveys conducted for shore fishing off docks, piers, jetties, breakwaters, bridges, causeways, beaches, and banks and for private, rental, and for-hire boats. These surveys collect data on (1) the angler by recording their state and county of residence and telephone number; (2) the trip by recording the state and county of trip, fishing mode, and area fished; and (3) the catch by recording the identified species, number of species, weight and length of landed fish, and disposition (i.e., thrown back dead or alive, used for bait, or kept to eat or sell). Total catch recorded for the intercept surveys include the landed catch as observed by the interviewer, and the catch landed and thrown back dead as reported by the angler.

Since 1996, aerial surveys have been conducted in Tampa Bay, Florida, by the Florida Fish and Wildlife Conservation Commission. This method provides instantaneous counts of boats engaged in fishing and their locations. Boats engaged in fishing are defined by observed fishing rods engaged in fishing and no commercial markings on the boat. Aerial counts are adjusted for "turnover" from the access-point or roving boat surveys to expand fishing effort estimates. Roving boat-access, shore roving-creel, and access-point surveys are used to correct effort for fishing guides and charter vessels not distinguished by the aerial observer.

FOR-HIRE SURVEY

FHS was first implemented in the Gulf of Mexico in 2000 but since has been extended to all coasts for all fisheries. This survey is designed to ascertain fishing effort and CPUE data. Effort is determined from boat directory telephone surveys, and CPUE is determined from access-point intercept surveys for charter and head boats and from at-sea surveys for head boats. Boat directory telephone surveys stratify charter and head boats. Samples are taken from the vessel telephone and address directory so that 10 percent are contacted randomly for each weekly vessel survey. The owner or operator of the vessel is contacted, and fishing effort is determined by recording the number of boat trips, number of anglers, and areas fished in that week. Boat directory telephone surveys are checked with dockside validation of boat trips in order to correct for trip reporting errors. Access-point and at-sea intercept surveys are also used to correct for trips by boats not in the directory. These methods of quality control aim to more accurately estimate total angler trips and mean catch per angler trip, which are used to determine total catch.

Intercept surveys are conducted similarly to the methods described in the MRFSS. At-sea sampling involves an observer onboard the charter or head boat for Florida, Alabama, the Atlantic coast, and California. The observer records the number of fish landed and released (alive or dead), species identification, and effort needed to catch those fish.

PARTY CHARTER SURVEY

The Party Charter Survey (PCS) is structured similarly to FHS, but PCS only includes fishing trips for party and charter boats off California. Fishing effort is determined from boat directory telephone surveys, and CPUE is determined from access-point intercept surveys and at-sea sampling. Intercept survey methods are described in detail in the MRFSS section, while boat directory telephone survey and at-sea methods are described in detail in the FHS section.

ALASKA SPORT FISH STATEWIDE HARVEST SURVEY

The Alaska Sport Fish Statewide Harvest Survey (SWHS) was implemented in 1977 to obtain statewide estimates of catch, location, and CPUE for each species. SWHS is a fishing household mail survey sent out using the angler license directory. The survey samples about 20 percent of the households with licensed anglers, and about 40 percent of the sampled households currently respond to the survey. The SWHS was initially used instead of the MRFSS because there were not many telephones in Alaska. In 2003, nearly 292,000 anglers that fished and over 50 percent of the households surveyed were not residents of Alaska.

CALIFORNIA RECREATIONAL FISHERIES SURVEY

The California Recreational Fisheries Survey (CRFS) was developed in response to concerns from resource managers and constituents in regards to groundfish management. CRFS is implemented under the Pacific coast's Recreational Fisheries Information Network (RecFIN) to determine monthly estimates for quota management for all California fisheries. Fishing effort is determined from angler directory telephone surveys, and CPUE is determined from access-point roving-creel, and access-point boat-trip intercept surveys; the survey type varies depending on the fishing method. Both the angler directory telephone and roving-creel surveys were initiated in California in 2004.

Angler directory telephone surveys randomly sample from the angler license telephone and address directory. The angler is contacted, and fishing effort is determined by recording the number of boat trips, number of anglers, and areas fished. Angler directory telephone surveys are checked with dockside validation of boat trips in order to correct for trip reporting errors. Access-point, roving-creel, and boat-trip intercept surveys also are used to correct for trips by anglers not in the directory. Similar to effort determined using the FHS, these methods of quality control aim to more accurately estimate the total number of angler trips and mean catch per angler trip, which are used to determine total catch.

Party and charter boats are sampled by a weekly telephone survey to determine effort. 10 percent of all active vessel skippers are surveyed to provide details on the number of trips for the week, trip type, and the number of anglers carried for fishing trips. Vessel operations are validated through field observations. Total catch; discards; area of catch for each stop with catch, depth, and length of discards; and angler demographics are determined from at-sea sampling.

Access-point boat-trip intercept surveys record data for private, rental, party, and charter boats. Public access sites are used to collect data from primary and secondary private boats.[1] Effort is determined by counting all primary boats returning to the site for the day, and CPUE is determined by sampling all boats at the completion of the trip and recording the number of anglers per boat, trip type, catch area, discards, identified catch, and weight and length of catch. The sites where secondary private boats are found are sampled as clusters. Effort is determined from instantaneous counts of boat trailers while roving the cluster of sites, and CPUE is determined by sampling individual anglers as boats return from the completion of the trip. Data recorded are similar to the CPUE data recorded for the primary private boat intercept surveys.

Private access sites, such as marinas, harbors, backyard slips, buoyed vessels, and private ramps and hoists, are used to sample private and

---

[1] Primary private boats are boats returning to sites where 90 percent of catch is recorded and catch includes management species of concern. Secondary private boats are boats returning to sites where 10 percent or less of catch is recorded.

rental boats. Sampling methods include onsite and offsite surveys (e.g., for night fishing). Voluntary catch logs are used by a panel of private access anglers and fishing clubs to determine CPUE. The logs are validated with public access boat ramps. Catch rates and CPUE from similar targeted trips at adjacent public launch ramps are included in private access boat creel data.

Shore sampling is conducted for fishing off of human-built platforms, which include piers, jetties, and breakwaters, and off beaches or banks. Fishing effort off of human-built structures is determined by counting the number of anglers at the beginning and end of the survey day and by tallying arrivals and departures during the day. Effort for beach and bank anglers is determined from angler license telephone surveys. CPUE for fishing off of human-built structures is determined by interviewing individual anglers at the completion of their trip. CPUE for beach and bank anglers is determined by roving creel access-point surveys. Identified catch, catch length and weight, discards, angler demographics, and license information are recorded for both human-built fishing structures and beach and bank fishing. Night and private-access fishing effort is estimated from licensed angler telephone surveys.

CATCH CARD SURVEY

The Catch Card Survey (CCS) is implemented in Washington, North Carolina, and Maryland. Catch effort is recorded by mandatory catch card reports regulated by the National Oceanic and Atmospheric Administration. The Washington survey includes all boats and shore fishing and covers the salmon and halibut fisheries. The Atlantic coast survey includes all boats and covers the bluefin tuna and marlin fisheries. CCS data only include the recreational landings of designated species. Highly migratory species catch cards provide a census for landed recreational billfish and Atlantic bluefin tuna.

LARGE PELAGIC SURVEY

LPS is used along the Atlantic coast from Virginia to Maine. LPS records information for fishing of offshore pelagic species within this region. Only private and charter boats with highly migratory species permits are sampled. Effort is determined from boat directory telephone surveys, and CPUE is determined from access-point intercept surveys.

Refer to the FHS section for details on boat directory telephone survey methods.

OREGON RECREATIONAL BOAT SURVEY

The Oregon Recreational Boat Survey (ORBS), which is applied for all fisheries in Oregon, is designed to sample ocean-going private, charter, and party boats. Effort for ORBS is determined from onsite boat exit count surveys. Private boats are surveyed by counting bar crossings from dawn to 10:00 AM in most ports, with expansion to include trips leaving outside of this time frame. Count surveys include the initial trailer and moorage slip count plus counts of additional launches throughout the day. An additional 4 percent expansion is made to all effort to account for late afternoon trips as recommended by the RecFIN Statistical Committee. Charter boat effort is calculated by contacting the charter offices for the tally of trips stratified by target species. This data is validated with bar crossing counts. CPUE is determined from port-based boat-trip intercept surveys. Private boats are interviewed throughout the major moorage and launch sites, and charter boats are interviewed usually with prior knowledge of trip type. Sampling is conducted independently of vessel size and passenger load. All interviews are conducted at the completion of the trip to tally catch by species retained, species released, length and weight for most species, lengths for Pacific halibut, and catch area.

Data are stratified by week and season type. Narrow time frames are required due to the highly variable season and because sampling rates can vary over the year. Pulse fisheries, like the deepwater halibut season (often Thursday through Saturday), require further stratification beyond the week level.

OCEAN SAMPLING PROGRAM

The Ocean Sampling Program (OSP) is administered for all fisheries in Washington and is designed to sample coastal private, charter, and party boats. Effort for OSP is determined from onsite boat entrance count surveys of vessels at ocean ports, and CPUE is determined from port-based boat-trip intercept surveys. Boat interviews are conducted at the completion of the trip to tally catch by species, catch length, catch area, and discards. Onsite boat interviews are described in detail in the ORBS section.

PUGET SOUND SAMPLING PROGRAM

The Puget Sound Sampling Program (PSSP) is used in Washington's Puget Sound to determine special area catch estimates and CPUE. PSSP methods are structured to cover the same site for the entire shift where site selection is based on anticipated effort. Therefore, PSSP data are responsive to pulse fisheries. Effort at each site is dependent on fishery openings and closings, catch success, and angler preference. Periodic spot checks are conducted to confirm effort expectations.

Effort is determined from telephone surveys, and CPUE is determined from boat-intercept surveys. The Washington Interactive License Database is used for the telephone surveys and electronically captures licensed angler contact information at the point of sale. Charter-issued licenses are outside the point of sale system. The angler license survey randomly samples 1,700 out of 200,000–300,000 licensed saltwater anglers every two months. The sampling frame is updated prior to each survey. The angler-license and charter boat-operator telephone surveys are conducted to collect trip-specific information. These data include the total number of trips, dates, marine catch area where fishing occurred, number of anglers per boat, launch sites, and target species. Response data and the number of contact attempts also are recorded. Adjustments for unlicensed anglers are estimated from intercept sampling. CPUE for PSSP is determined from the calculated mean catch per trip. Boat interviews are conducted at the completion of the trip to tally catch by species, catch length, catch area, and discards. The numbers of anglers per boat, licensed anglers per boat, and license type also are recorded.

SHORE AND ESTUARY BOAT SURVEY

The Shore and Estuary Boat Survey (SEBS) is implemented for all fisheries in Oregon. Shore and inland boats are surveyed using an angler-license frame telephone survey to determine effort and an access-point intercept survey to determine CPUE. Ocean-going private, charter, and party boats are surveyed using onsite boat exit count surveys to determine effort and port-based boat-trip intercept surveys to determine

CPUE. Species caught, catch length and weight, catch area, and discards are collected to calculate CPUE. Details of the SEBS survey methods are described above in various sections.

SOUTHEAST HEAD BOAT SURVEY

The Southeast Head Boat Survey is used for all fisheries in the Gulf of Mexico. Effort, CPUE, and catch for head boats are determined from logbook census. The logbooks are collected dockside every two weeks, on average, by the port agent. At times, such as the off-season, logbooks are mailed in by the boat captain. This onsite collection is one way that the information is verified. Additional verification is gained through onsite surveys that are done at the end of trips to gather sampling data to compare to the logbooks. The logbooks are used to gather boat permit number and identification details, date and time sailed, area sailed (e.g., state waters, federal waters, inshore), length of trip, number of anglers, catch by species, catch location (which is done in a 10 minute by 10 minute grid), and discards. (Discard data are separated out into living and dead categories.)

TEXAS MARINE RECREATIONAL FISHING SURVEY

The Texas Marine Recreational Fishing Survey was initiated in 1974 and is structured to collect information from private, rental, and charter boats regarding the targeted species, catch composition, catch number, and catch size through stratified proportional random sampling. Data on trip length, angler CPUE, location of fishing, gear and bait used, residence of anglers, and trip satisfaction also are collected.

Onsite surveys are conducted to collect trip-specific information, and roving surveys are done to collect trailer and empty wet-slip counts. Results from the onsite survey are expanded by relative pressure at that site. Boat access sites are surveyed in relation to fishing pressure. Surveys are conducted for 1,000–1,800 hours to maximize angler intercept. Fishing seasons are stratified by high use (May 15–November 20) and low use (November 21–May 14). This is stratified further by day type (weekdays, weekends, and holidays). Surveys are conducted 97 days per bay during high-use season and 36 days per bay during low-use season, except for Sabine and San Antonio, which only have 72 high-use survey days. Two-thirds of the surveys are conducted on weekdays and one-third on weekends. This totals to an annual coastwide sample of 1,014 survey days and approximately 12,000 fishing trips interviewed per year. Further details of onsite surveys are described above in the CRFS section.

VESSEL TRIP REPORTS

Vessel Trip Reports (VTRs) are implemented for federally regulated fisheries off the Atlantic coast. Effort, CPUE, and catch for party, charter, and head boats with federal permits are determined from logbook census. The logbooks, which are collected from these boats for each trip in state or federal waters, are required to be submitted by the fifteenth of the month for all trips in the previous month. VTRs record the boat permit number and identification details, date and time sailed, trip type (i.e., party or charter), number of crew, number of anglers, catch by species, catch location, and discards. (Discard data are not separated out into living and dead categories.) Currently, there is no dual-frame system in place to verify the information given in VTRs.

Appendix C

Fisheries Case Studies

This appendix discusses case histories for three important recreational fisheries in three different U.S. regions. These fisheries were chosen because they represent a range of scale and management intensity. (These histories were drawn from information available during the summer of 2005; therefore, they may no longer represent the current conditions. Nonetheless, these examples are useful to the larger discussion.) The case history for lingcod (*Ophiodon elongatus*) represents a large-scale, federal stock assessment with coastwide management goals set by the Pacific Fishery Management Council (PFMC); however, in-season management is conducted by individual state agencies. The case history for red snapper (*Lutjanus campechanus*) in the Gulf of Mexico represents a large-scale fishery conducted in federal waters. Although management goals are annual, in-season management has failed in this fishery, and the Gulf of Mexico Fishery Management Council (GMFMC) actually evaluates management on the same schedule as the stock assessment—every three to five years. The case history for striped bass (*Morone saxatilis*) represents an east coast fishery conducted primarily in state waters with a coastwide stock assessment and annual management objectives set by a regional advisory board with individual states overseeing annual management.

LINGCOD

Life History

The lingcod (Figure C.1) is the largest member of the greenling family. They occur from Kodiak Island, Alaska, to Baja California, Mexico, in depths to 475 meters (m) (1,558 feet [ft]) but usually shallower than 300 m (984 ft) in rocky habitats and kelp beds. They attain a maximum length of about 1.5 m (5 ft) and can weigh over 31.8 kilograms (kg) (70 pounds [lbs]). They are voracious predators on fish, shellfish, and octopus. Lingcod are considered non-migratory; although, tag data show some individuals may move great distances. Males and females tend to be separated by depth, with females preferring deeper water. Males guard egg clutches ("nests") until hatching, generally 7–11 weeks. During this period, the males are territorial and very susceptible to harvest. If the males are removed from the nest, other organisms consume the eggs. The maximum published age is 20; however, lingcod have been aged to 26 in Alaska.[1] Richards et al. (1990) examine length and maturity relationships of lingcod from three areas in British Columbia and find that males begin to mature at 0.50 m (1.64 ft) and are all mature at 0.70 m (2.30 ft), and females begin to mature at 0.50 m (1.64 ft) and are all mature at 0.75 m (2.46 ft).

Current Stock Status and Management Authority

Lingcod along the U.S. west coast are managed as two distinct stocks: the Lingcod-North stock (Washington and Oregon) and the Lingcod-South stock (California). (Alaskan lingcod are not considered in this case study.) Both stocks are listed as overfished and are subject to a rebuilding plan. The latest assessment indicates that the lingcod stock has achieved its rebuilding objective of $B_{40\%}$[2] in the north (actually 28 percent above $B_{40\%}$) but was at $B_{31\%}$ in the south. PFMC sets quotas

---

[1] Personal communication, Kristen Munk, Alaska Department of Fish and Game, Juneau.

[2] B represents the stock biomass (weight of a population of fish). The subscript represents the percent of the stock relative to its unfished biomass. $B_{40\%}$, a proxy for B_{MSY}, is the biomass needed to sustain maximum sustainable yield (40 percent of the unfished biomass).

FIGURE C.1 A young deckhand with a tagged lingcod (photo used with permission from Charlie Wilber).

for lingcod fisheries, with in-season management by the individual states.

Fishery Characteristics

There are both recreational and commercial fisheries for lingcod, with recreational being particularly important in the southern area. In 2002, anglers landed 577 metric tons (mt) (1.3 million lbs) out of a total optimum yield (OY) of 577 mt (1.3 million lbs) and a total postseason catch estimated at 779 mt (1.7 million lbs) (51 percent over the coastwide OY). Private vessels and rental vessels accounted for most of the recreational catch, with the majority of the catch coming from Oregon and northern California. Charter and party boats accounted for 7 percent of recreational catch in Washington, 43 percent in Oregon, 32 percent in northern California, and 14 percent in southern California. Commercial lingcod fisheries along the west coast have been predominately trawl fisheries. Recently, restrictions in trawl fisheries have increased; therefore, roughly half of the commercial landings are now from hook and line. Coastwide commercial landings totaled 223 mt (492,000 lbs) in 2002 (Jagielo et al., 2003).

Since 2004, PFMC has set two separate recreational fishery quotas for the two stocks. In 2003, California exceeded the entire west coast limit for lingcod that resulted in a coastwide late-season closure. Consequently, in 2004, PFMC set separate state fishery targets (Pacific Fishery Management Council, 2004); California is given statewide recreational allocation, and Oregon and Washington are given a combined recreational allocation. The recreational sector took 63 percent of the total catch, and the California recreational fishery took 61 percent of the coastwide recreational catch in 2004. In 2005, the recreational catch guidelines were 422 mt (930,000 lbs) for California, 151 mt (333,000 lbs) for Oregon, and 83 mt (183,000 lbs) for Washington. Size limits, bag limits, depth limits, and seasons vary by state and within state; in-season management action may be taken to prevent exceeding annual quotas.

Oregon has committed to updating anglers on allowable catches on a monthly basis during the season so data on catch are required in a timely fashion. In-season changes are made in July if it appears that current catch rates will cause an in-season closure.

Recreational Survey Methods

The recreational fishery survey for lingcod is multifaceted. The Recreational Fisheries Information Network (RecFIN) program stores the survey data, the Marine Recreational Fisheries Statistics Survey (MRFSS) is used for the historical California data, field-intercept surveys are administered by the Pacific States Marine Fisheries Commission, and a random telephone survey of the coastal population is administered by a National Marine Fisheries Service (NMFS) contract. Additionally, there is the California Charter and Party Fishing Vessel effort survey (since 2001), the California Recreational Fisheries Survey (CRFS) (replaced the MRFSS in 2004), the Oregon Recreational Boat Survey, and Washington State's Ocean Sampling Program (OSP). The Oregon Recreational Boat Survey includes a field-intercept survey for effort and catch of private and rental boats and party and charter boats and a telephone survey of license holders for shore and estuary boat anglers, and OSP includes a seasonal exit count of vessels and an intercept sampling of catch.

Stock Assessment Method

The catch-at-age model used is a multiple-fleet, age-, and sex-structured model implemented in Coleraine (Jagielo et al., 2003). Coleraine is a general age-structured model used for fish-stock assessment developed by Hilborn and colleagues; a manual and other information are available through the University of Washington (2004). Several weaknesses with the model include the lack of a recreational fishery catch per unit effort (CPUE) index. Although recreational catch accounts for most of the removals, recreational CPUE is problematic because catch rates may be affected by variable target species, undocumented search time, unreported discards, unknown spatial effort shifts, and bag limit effects (Jagielo et al., 2003). These problems will be exacerbated by changes in management, such as depth and seasonal closures. Information on discard and discard mortality is difficult to estimate but is important to the model. According to Jagielo et al. (2003):

> MRFSS has collected B1 (reported by angler to be dead) and B2 (reported by angler to be alive) catches since 1980. Estimates of lingcod discarded alive have increased substantially in response to (1) management changes in 1998 (the size limit increased from 22 to 24 inches) and (2) a seasonal closure in California waters beginning in 2000… It is interesting to note that estimates of fish discarded dead have decreased over time. Estimated live lingcod discarded in southern California was 306,000 fish in 2002. This compares to a total landed catch of 25,000 fish. [Washington Department of Fish and Wildlife] began collecting discard information from the recreational fishery in 2002 and estimated that 57 percent of the catch was discarded. [Washington Department of Fish and Wildlife] does not collect information on the portion of the catch discarded live or dead. Based on an earlier study…, the PFMC Groundfish Management Team used a 20 percent inflation factor to adjust landed catch to account for unobserved lingcod mortality in the commercial fishery beginning in 2002. Data collected by the Groundfish Observer program in 2001–2002 estimated that the percent discard of total observed catch was 78.8 percent. Because lingcod lack a

> swim bladder, it is likely that there is a relatively good survival rate for these fish.

In the models projecting future catch, both Washington and Oregon use 5 percent mortality for live discards, but this mortality is not incorporated into the current year assessment. There have been no formal studies of lingcod hooking mortality.

Removals

Recreational catch data for the lingcod assessment come from a variety of sources. For California, the RecFIN database (including the MRFSS) was used for 1980–1989 and 1993–2003. Oregon catch data were provided by the Oregon Department of Fish and Wildlife. Washington catch data were obtained from the Washington Department of Fish and Wildlife (WDFW) OSP. Beginning in 2004, CRFS has been used in place of the MRFSS for California. Commercial catch data were compiled from agency reports and personal communication for all years preceding 1981. The Pacific Coast Fisheries Information Network (PacFIN) database was queried for catch information in subsequent years.

Demographics

The Lingcod-North population age data are available for recreational fisheries for 1980 and 1986–2002 with sample sizes ranging from 226 to 1098. The Lingcod-South population age data are available for recreational fisheries for 1992–1998 and 2000–2002 with sample sizes ranging from 48 to 545. Weight is estimated using a von Bertalanffy growth equation, which is updated periodically. Other sources used for size and age data include the commercial fisheries, the triennial trawl shelf survey, and the WDFW Cape Flattery Tag survey. The Stock Assessment Review Panel (STAR Panel) suggested that more emphasis needs to be placed on collecting biological data to improve fishery age, length, and sex samples sizes and to improve geographic coverage (Pacific Fishery Management Council, 2003).

Abundance Indices

The 2003 stock assessment for west coast lingcod uses commercial fishery and fishery-independent indices of abundance, such as the following:

- NMFS Triennial Shelf Trawl Survey (biomass and associated coefficients of variation) (This survey does not cover rocky habitat, a prime lingcod habitat and therefore has a habitat bias.)
- WDFW Cape Flattery Tag Survey (length composition used as a recruitment index)
- Trawl fishery logbook CPUE index

Another index considered but not used is recreational CPUE, which was not used because of high index variability, lack of a discernable index trend, implausible temporal changes in abundance, and unresolved input data assumptions. Data from Washington did not contain discard information so there was no way to convert its data to total catch, which is important in estimating a CPUE trend. Also, Jagielo et al. (2003) find that "recreational CPUE data sets are often problematic for use as unbiased indices of abundance because catch rates may be affected by (1) variable target species by boat, (2) undocumented search time, (3) unreported discards, (4) unknown spatial effort shifts, and (5) bag limit effects." Because the recreational fishery now takes over 70 percent of the total catch of lingcod, the lack of an abundance index for this sector will continue to be problematic.

Recommendations for Data Collection

Jagielo et al. (2003) provide the following recommendations regarding data collection for this resource:

- Improve fishery age-structure sampling size and geographic coverage
- Conduct more frequent and synoptic fishery-independent surveys for stock indices and recruitment index
- Analyze CPUE on a reef-specific basis for evaluation of an index of abundance

- Enumerate at-sea discards and mortality of released recreational fish coastwide to account for total mortality

Economic and Social Information

Commercial landings of lingcod have been reduced greatly in recent years by management actions. In 2004, 166 mt (366,000 lbs) of commercial lingcod were landed on the west coast for an ex-vessel (dockside) value of $412,000. No data are specifically available on the economic value of the recreational fishery for lingcod. In 2003, there were 1,283 angler trips directed for groundfish on the west coast and, of these, 325 were charter trips, and 958 were private vessel trips.

RED SNAPPER

Life History

Red snapper (Figure C.2) are found along the Atlantic coast of North America from North Carolina to the Florida Keys and along the Gulf of Mexico from Florida to the Yucatan peninsula of Mexico (Robins et al., 1986). Adults are found in submarine gullies and depressions; over coral reefs, rock outcrops, and gravel bottoms; and are associated with oil rigs and other artificial structures (Gulf of Mexico Fishery Management Council, 2003). Eggs and larvae are pelagic while juveniles are found associated with bottom features or over barren bottom. Spawning occurs during the summer and fall over firm sand bottom with little relief away from reefs. Adult females mature as early as two years, and most are mature by four years (Schirripa and Legault, 1999). Red snapper have been aged up to 53 years, but most caught by the directed fishery are 2 to 4 years old (Wilson and Nieland, 2001).

Current Stock Status and Management Authority

The red snapper stock is in an overfished condition and continues to undergo overfishing (Gulf of Mexico Fishery Management Council, 2005a). This stock has been overfished since at least 1988. Currently,

FIGURE C.2 Anglers with their red snapper (photo used with permission from Jon G. Sutinen).

it is under a rebuilding plan to end overfishing in 2009–2010 and to rebuild the stock to B_{MSY} by 2032 (Gulf of Mexico Fishery Management Council, 2004a). The current population is now dominated by young fish, creating a low spawning potential ratio. In addition to the mortality associated with shrimp trawl bycatch, the population shows signs of overharvest in the directed fisheries, with truncated ages.[3]

GMFMC establishes management plans and regulates the recreational red snapper fishery in U.S. federal waters. In the Gulf of Mexico, bag and minimum size limits and area and season closures are used. Individual states generally set regulations in state waters to comply with the federal regulations. However, Texas has a different recreational size limit and does not have a closed season to recreational fishing. Florida opens the recreational red snapper fishing season six days earlier than in federal waters.

---

[3] Personal communication, Felicia Coleman, Florida State University, Tallahassee.

Fishery Characteristics

The red snapper fishery occurs predominately in federal waters. Red snapper have supported an important commercial fishery in the Gulf of Mexico for more than a century, with 6,400 mt (14.1 million lbs) landed in 1900. Documentation of the recreational fishery began on a regular basis in 1981 with the MRFSS.

According to Schirripa (1999):

> Management of the red snapper resource has meant dividing the allowable fishing mortality between two competing fisheries: the directed fishery, which consists of a commercial and a recreational sector, and the undirected shrimp fishery. The shrimp fleet harvests age 0 and 1 red snapper in the form of bycatch. These vessels use bottom trawls to harvest shrimp, which share a propensity for the same habitat as juvenile red snapper. Although the discarded catch associated with shrimp trawls is not counted toward the [total allowable catch], it is included in the stock assessment as part of the total fishing mortality. In 1991, turtle excluder devices (TED) were mandated for all offshore shrimp boats operating in the Gulf of Mexico. In 1998 all offshore shrimp boats fishing in the western Gulf of Mexico were required to use some form of bycatch reduction device (BRD) as well. Several types of BRDs have been "certified" by the U.S. fishery management authorities as reducing red snapper bycatch by 30 percent to 50 percent with an approximate 4 percent to 6 percent reduction in shrimp loss.

However, at present, BRDs are estimated to reduce red snapper bycatch by only 11.7 percent (Foster, 2004). Further reductions in overall red snapper bycatch may have occurred as a result of reductions in shrimping effort due to the depressed economic condition of the shrimp fishery and due to the loss of shrimp vessels and processors from Hurricanes Katrina and Rita in 2005.[4]

Red snapper is the most popular offshore recreational finfish fishery in the northern and western Gulf of Mexico with 49 percent of the di-

---

[4] Personal communication, Steven Atran, GMFMC, Tampa, Florida.

rected fishery's total allowable catch (2,041 mt [4.5 million lbs] in 2005) allocated to the recreational sector. Fishing occurs off the coasts of Florida, Alabama, Mississippi, Louisiana, and Texas. More than 70 percent of the recreational catch is taken by the for-hire sector (Gulf of Mexico Fishery Management Council, 2004a). In addition to landed catch, more than half of all recreationally caught red snapper are released because of regulatory limits; release mortality for the recreational fishery is estimated to range from 15 to 40 percent (Gulf of Mexico Fishery Management Council, 2004b; Table C.1). Red snapper is listed as overfished; therefore, staying within the total allowable catch is particularly important (Gulf of Mexico Fishery Management Council, 2005a,b). During the 1990s and early 2000s, the recreational sector exceeded its allocation 10 out of 12 years, often significantly (Gulf of Mexico Fishery Management Council, 2004b). In response to such frequent and large overruns, Congress, in the Sustainable Fisheries Act of 1996, mandated that the recreational red snapper fishery in the Gulf of Mexico be managed in-season with a quota and that the fishery be closed when the quota is reached.

In-season management was attempted during 1997–1999 using extrapolation of available data (Gulf of Mexico Fishery Management Council, 2005b). During this period, the annual recreational catch was based on the MRFSS, the Texas Parks and Wildlife Department's (TPWD) coastal sport fishing survey, and NMFS's head boat survey; discard information was collected in the MRFSS only. In-season management was difficult because the recreational catch data are not available for several months after collection. NMFS made projections of the recreational catch using average catch data from the previous two years from all three surveys. The data from the surveys done by Texas and NMFS were not available in-season. When available, the MRFSS data from the first two months of the current year replaced the projection for those two months, and this continued as current data became available but was limited to the first six months of the year; the last six months were estimated using previous years' data (Gulf of Mexico Fishery Management Council, 2005a). The recreational fishery was closed in-season with three weeks notice on November 27, 1997; September 30, 1998; and August 29, 1999; when the recreational sector was projected to have reached its share of the total allowable catch. These projections were not accurate, and the recreational sector overharvested its allocation by 20 percent in 1997, 29 percent in 1998, and 23 percent in 1999. In 2000, GMFMC abandoned in-season management of the recreational sector and moved to a shorter season, larger

TABLE C.1 Summary of Depths Fished and Estimate of Release Mortality by Fishery and by Region in the Gulf of Mexico

Recreational Fishery	Depth in m (ft in parentheses)	Percentage of Release Mortality
Eastern Gulf	20–40 (66–131)	15.0
Western Gulf	40 (131)	40.0
Gulfwide (1981–1996)		27.5
Gulfwide (1997–2002)		21.0
Commercial Fishery—Open Season		
Eastern Gulf	55 (180)	71.0
Western Gulf	58 (190)	82.0
Gulfwide (1962–1983)		73.0
Gulfwide (1984–1992)		77.0
Gulfwide (1993–2003)		80.0
Commercial Fishery—Closed Season		
Eastern Gulf	55 (180)	71.0
Western Gulf	83 (272)	88.0

NOTE: Release mortality is expressed as the percent of discarded fish that were assumed to suffer mortality.

SOURCE: Gulf of Mexico Fishery Management Council, 2004b.

size limits, and smaller bag limits to try to keep the recreational catch within its allocation. GMFMC has requested more timely and accurate reporting of recreational catch data to allow annual management.

The quota for recreational and commercial fisheries is set annually and is not to be exceeded on an annual basis; however, recreational catch data is compiled only when periodic stock assessments are done (generally every three to five years), and changes to regulations are made accordingly. Many public and agency comments were critical of in-season changes in size, bag limits, and seasons for the recreational sector, favoring stability in regulations. The current quota is 4,128 mt (9.10 million lbs), 2,105 mt (4.64 million lbs) to commercial and 2,023 mt (4.46 million lbs) to recreational. The directed commercial season begins February 1 at noon and closes February 10 at noon. The season runs for the first 10 days of each subsequent month until two-thirds (1,406 mt

[3.1 million lbs]) of the commercial quota is reached. The remaining one-third (680 mt [1.5 million lbs]) is landed in the fall, and there is a minimum size limit of 0.4 m (15 inches [in]). The recreational season runs from April 21 through October 31, and the recreational bag limit is four fish with a minimum size limit of 0.4 m (16 in).

Currently, a vessel moratorium is under way for charter and head boats as this sector takes the largest share of the recreational catch. An individual fishing quota program is being considered for the directed commercial fishery.

Recreational Survey Methods

Recreational survey methods are diverse for this fishery and include a shore-based, private boat, rental boat, and for-hire fishing mode that uses the MRFSS coastal household telephone survey and the MRFSS intercept survey. The for-hire fishing mode is surveyed using the Vessel Frame Telephone Survey (prior to 2006, this was the Southeast Head Boat Survey) and the Vessel Effort Validation Survey. The MRFSS access-point intercept survey is used for sampling catch. Additionally, TPWD's coastal sport fishing survey provides estimates for numbers harvested by boat modes exclusive of party boats for Texas for 1986 and later years. Harvest by shore-bound anglers has not been included in the Texas estimates since 1985. This survey covers the private, rental, and charter boat mode for this state. However, the majority of recreational landings of red snapper from Texas come from the head boat fishery, which was included in NMFS's head boat survey.

Stock Assessment Method

The red snapper assessment now uses an age-structured stock assessment model called CATCHEM, which is a two-stock model with eastern and western stocks (Gulf of Mexico Fishery Management Council, 2004b). An age-structured analysis program (ASAP) single-stock model was used prior to 2005 (Legault and Restrepo, 1998). The ASAP model synthesized a variety of available data to develop reference points that best fit the behavior of a simulated age-structured population. CATCHEM is a complex model that relies on both historic and current data from all fishing sectors (Porch, 2004a, b). The CATCHEM model was developed as an alternative to the two-step approach used in prior assessments where the number of fish discarded owing to minimum size limits was determined by the probabilistic method of Goodyear (1997) and then used along with indices of abundance in an age-structured model (ASAP or Virtual Population Analysis [VPA]). The discards from the recreational and commercial sectors during the open season were assumed to occur predominantly due to the regulations on minimum size. They were computed on a seasonal rather than annual basis to better accommodate the rapid growth exhibited by younger red snapper. Recreational catch data, recreational effort data, and age data from recreational samples are all inputs into this model. Since 1981, the annual recreational catch and effort estimates have been based on the MRFSS, TPWD's coastal sport fishing survey, and the NMFS head boat survey.

Removals

As stated in the Southeast Data, Assessment, and Review (SEDAR):

> Commercial landings statistics are the quantities and value of seafood products sold to established (licensed) wholesale and retail seafood dealers. Currently, these data are collected by trip ticket programs managed by the state fishery agencies in Florida, Alabama, and Louisiana [Gulf of Mexico Fishery Management Council, 2004b]. Dealers in Mississippi and Texas are required to submit monthly reports that provide quantity and value by species. Prior to the implementation of the trip ticket programs, landings statistics were collected by [NMFS] and state employees that visited the seafood dealers monthly and recorded the quantities and value purchased for each species for a calendar month. In addition, the agents would assign an estimate of the type of gear and fishing area where the landings were caught.
>
> The Southeast Fisheries Science Center (SEFSC) has maintained the commercial landings statistics (also known as general canvass landings statistics) in a regional database since the mid-1980s. The states provide the landings statistics from their trip ticket or monthly program to the SEFSC, and these data are summarized

> and maintained in the same format as the historical general canvass data. (Gulf of Mexico Fishery Management Council, 2004b)

Directed commercial fishery discards are recorded on discard forms provided as part of a mandatory logbook requirement.

Recreational data were compiled for the three following sources:

- The MRFSS (1981–1998) with some exceptions: (1) no wave 1 data in 1981, (2) no Texas boat mode in 1982–1984, (3) no Texas data after 1986, and (4) no head boat sampling after 1985
- NMFS's Beaufort Laboratory head boat survey for all states after 1985
- TPWD's coastal sport fishing survey

Data on recreational discards were collected by the MRFSS but were not available for Texas landings or for landings from head boats. Red snapper catches from Texas only account for about 2 percent of the annual recreational landings gulfwide.

Demographics

According to Schirripa and Legualt (1999):

> Morphometric, growth, and other biological characteristics of red snapper were evaluated using a composite of length and other measurements of Gulf of Mexico red snapper that have been collected during research and monitoring programs through the years. The present evaluation combined the data from prior analyses with more recent observations from a variety of sources. A description of the earlier data and sources are given in Parrack (1986a, b) and Parrack and McClellan (1986), who obtained the data and prepared computer files of the various data sets. In addition, data collected during the trip intercept portions of the [MRFSS]; the [NMFS] head boat survey; and samples of commercial and recreational catches collected as part of the Trip Interview Program (TIP) of the State/Federal Cooperative Statistics Pro-

> gram provided additional data sources. A biological profiles sampling program by the [NMFS] Panama City (Florida) Laboratory provided additional observations of growth and fecundity, as well as morphometrics. Additional data were provided from research programs at the University of South Alabama, Louisiana State University, the University of West Florida, and the Louisiana Universities Marine Consortium.

Currently, there are only very limited data on sizes of live discards from the recreational sector. This is important because 60 percent of recreational catches are discarded gulfwide (Gulf of Mexico Fishery Management Council, 2004b).

Abundance Indices

Abundance indices include four fishery-dependent sources and three fishery-independent sources. The fishery-dependent indices are the commercial handline logbook data (directed and bycatch), including the eastern and western Gulf jig fishery and bottom longline data; bycatch of red snapper in shrimp trawl fishery (pre-1990 when bycatch could be sold); the MRFSS and TPWD's coastal sport fishing survey's CPUE data; and the head boat catch rates from the NMFS head boat survey. The fishery-independent indices are the Spatial Ecological Analysis of Megavertebrate Populations (SEAMAP) database for shrimp and bottomfish, for ichthyoplankton (under consideration as an index), and for reef fish (under consideration as an index).

Economic and Social Information

Red snapper is the most valuable commercial reef fish fishery in the Gulf of Mexico, with dockside landings worth over $9 million in 1998. Individual fishing quotas are being considered in the directed commercial fishery.

Red snapper is also a very important recreational species in the Gulf of Mexico. Shifts in the recreational sector have occurred over the past two decades; in 1981 and 1982, private anglers landed about 65 percent of the recreational red snapper reported. However, presently, the for-hire sector lands over 70 percent of the recreational catch, with charter vessels taking most of this catch (Gulf of Mexico Fishery Management Council, 2004b). There are an estimated 3,220 recreational for-hire vessels in the Gulf of Mexico (Federal Register, 2002). GMFMC began notice of a federal for-hire sector moratorium in November 1998. Initial implementation occurred in 2001, but due to an error in the public notice, the application process was reopened in 2003 and again in 2005. The intent is to limit the for-hire sector to the March 29, 2001, participation level. Currently, approximately 1,554 vessels are licensed to participate in the moratorium for reef fish (Gulf of Mexico Fishery Management Council, 2005c). Compliance with the federal licensing requirement has increased with the development of the moratorium.

STRIPED BASS

Life History

The Atlantic striped bass (Figure C.3) is a migratory species that ranges from the St. Lawrence River in Canada to the St. John's River in Florida. (The west coast population of striped bass, resulting from a few hundred fish introduced from the Navesink River in New Jersey to San Francisco Bay in the 1880s and now occurring from southern California to British Columbia, is not considered in this discussion.) This species can live up to 30 years and spends most of its adult life either in coastal estuaries or in the ocean, migrating north in the summer and south in the winter. Striped bass are anadromous, and in the spring, adults ascend rivers to spawn. Along the U.S. Atlantic coast, the major part of the migratory stock originates in the Chesapeake Bay spawning areas, with significant contributions from spawning grounds in the Hudson and Delaware Rivers. Fertilized eggs and larvae drift downstream to nursery areas in river deltas, inland portions of coastal sounds, and estuaries where they mature into juveniles. The juveniles remain in these areas for two to four years and then join the coastal migratory population. Females are highly fecund producing 0.5 million eggs at six years of age and 3 million at 15 years of age.

FIGURE C.3 Striped bass caught by young anglers in Maryland's Chesapeake Bay (photo used with permission from the National Oceanic and Atmospheric Administration).

Current Stock Status and Management Authority

The striped bass population has been increasing steadily since 1982 and, in 2004, the population was estimated to be 11 million fish higher than the average stock size for the previous five years and 23.8 percent higher than the population in 2003. The 2003 year-class was estimated at 22 million fish at age 1 and is the largest year-class in the time series (1982–2003). The most recent full stock assessment (Northeast Fisheries Science Center, 2003) determined that, in 2002, the stock was not overfished and that overfishing did not occur. However, the Atlantic Striped Bass Technical Committee reported difficulty in determining if overfishing was occurring in 2003 because of divergent patterns in fishing mortality estimates from VPA and analysis of tagging data. The next scheduled full stock assessment will be in 2007.

The implementation of the fishery management plan is mandatory under the Atlantic Striped Bass Conservation Act (P.L. 98-613). The Atlantic States Marine Fisheries Commission's (ASMFC) Striped Bass Management Board and Striped Bass Plan Review Team are responsible for monitoring the implementation of the fishery management plan. States are granted flexibility to deviate from the standards in the fishery management plan by submitting proposals for review by the ASMFC Striped Bass Technical Committee and Advisory Panel with approval from the ASMFC Management Board.

Fishery Characteristics

The fishery is limited to state waters (no catch allowed in federal waters). Much of the catch occurs in estuarine waters, but the MRFSS is limited to saltwater and, thus, not all estuarine landings are surveyed. In 2003, the total catch, including landings and discards, was estimated to be 4.7 million fish. The 2003 catch was above the 1996–2003 average of 4.0 million. Catch (2.4 million fish) and discards (1.2 million fish) from the recreational sector accounted for 76 percent of the total 2003 catch. Private and rental boats account for 80 percent of the recreational catch. Maryland accounted for the largest portion of the recreational fishery with 21.8 percent of total recreational landings, followed by Massachusetts (16.9 percent), Virginia (16.7 percent), New Jersey (16.3 percent), and New York (13 percent). The remaining states each landed 5 percent or less of the total recreational landings. In 2004, Massachusetts took 17 percent of the landed catch, followed by North Carolina and Maryland (Munger et al., 2005).

Commercial catch (0.86 million fish) and discards (0.27 million fish) accounted for 24 percent of the total 2003 catch. Maryland commercial fisheries caught 50.8 percent of the total commercial landings, followed by Virginia (18.7 percent), the Potomac River Fisheries Commission (9.6 percent), New York (7.9 percent), and Massachusetts (6.4 percent). The remaining states each accounted for 4 percent or less of the total commercial landings.

As of January 1, 2004, all states are required to implement a two-fish bag limit with a minimum size of at least 0.7 m (28 in) for their recreational fisheries. Chesapeake Bay fisheries, Albemarle–Roanoke fisheries, and states with approved conservation equivalency proposals are exempt from these rules. The first two areas have more conservative fishing mortality targets than those set by ASMFC and are allowed to set their own seasons, harvest caps, bag limits, and size limits as long as their total catches stay below their targets.

Recreational Survey Methods

The MRFSS is used as a basis survey in most states. The shore and private and rental boat fishing modes are surveyed using the MRFSS coastal household telephone survey and the MRFSS intercept survey. The for-hire sector is surveyed using the Vessel Frame Telephone Survey, the Vessel Effort Validation Survey, and the MRFSS access-point intercept survey.

Jurisdictions with significant recreational fisheries (Massachusetts, Rhode Island, Connecticut, New York, New Jersey, Maryland, Virginia, and the Potomac River Fisheries Commission) are required to follow specific guidelines for supplementing the MRFSS collection of catch composition data and catch and effort information from these fisheries to achieve a 20 percent coefficient of variation, or propose specialized striped bass surveys to better assess recreational catch.

Stock Assessment Method

The striped bass population dynamics are modeled using an ADAPT–VPA[5] model and an integrated catch-at-age model. Both models consist of a sequential population model which incorporates catch-at-age time-series data and research vessel survey information. The major differences between the two methods are that the integrated catch-at-age model assumes that the catch-at-age data are measured with error, and separability patterns are estimated as parameters. Data from eight tagging programs, which are conducted by the U.S. Fish and Wildlife Service Cooperative Striped Bass Tagging Program and have been in progress for at least 11 years, were used to provide alternate estimates of fishing mortality.

[5] ADAPT–VPA models are used extensively for fisheries stock assessments.

Removals

Given that the recreational fishery accounts for the largest portion of the striped bass catches, the MRFSS data are essential for accounting for total removals and for developing the catch-at-age information for the stock assessment. Total landings from the recreational striped bass fishery are calculated using type A+B1 records from the MRFSS. In 2003, these landings were estimated to be 2.4 million fish or 11,486 mt (25.3 million lb), with 80 percent accounted for by the private and rental mode in the MRFSS. Type B2 catch was estimated at 14.6 million fish. Size-at-age information on these discards indicate that it is the smaller fish that are generally released (peak at age 3; age range is 1–13 years). An 8 percent hooking mortality rate is assumed for recreational discards, resulting in an estimated loss of 1.2 million fish in 2003. Diodati and Richards (1996) examined mortality of striped bass hooked and released in saltwater. Predicted annual mortality averaged 9 percent and ranged from 3 percent under ideal conditions to 26 percent for the worst set of conditions. They found that surviving hooked fish had significantly lower physical condition factors than fish that had not been hooked.

The 2004 Striped Bass Stock Assessment Technical Committee expressed concern that there was considerable error in the catch produced by the MRFSS in 2003. The MRFSS estimated increases in some of the waves even though fishing effort was reported to have decreased due to hurricanes. On the other hand, there was also concern that the harvest had been underestimated because the winter fisheries (wave 1) in North Carolina and Virginia were not covered.

Estimates of commercial discards for striped bass rely on direct measurements from fisheries in the Hudson River Estuary and Delaware Bay and River and tagging information for the other areas. Since 1982, estimates for the other areas have been based on the ratio of tags reported from the discarded fish in the commercial fishery to tags reported from discarded fish in the recreational fishery, scaled by total recreational discards.

Demographics

Length and weight measurements usually are recorded by intercept interviewers for type A fish catch records. These data have been used to develop catch-at-length and catch-at-age estimates. Length frequencies of recreational landings for striped bass were based on a combination of the MRFSS length samples and volunteer angler logbooks. The age compositions of the recreational catch for each state were estimated using state specific age–length keys. Lengths from the volunteer angler logbooks and the American Littoral Society data were used to estimate the age composition of the recreational discards.

For the striped bass assessment, mean weights-at-age in the 2003 catch were determined from Maine and New Hampshire recreational harvest and discards; Massachusetts recreational and commercial catch; Rhode Island recreational and commercial catch; Connecticut recreational catch; New York recreational catch and commercial landings; New Jersey recreational catch; and Delaware, Maryland, Virginia, and North Carolina recreational and commercial catches.

Abundance Indices

Indices of abundance for striped bass represent both fishery-independent and fishery-dependent sources. There are seven sets of fishery-independent indices used for the adult population: the Maryland gillnet survey of the spawning population (ages 2–13+), Virginia pound net CPUE (ages 2–13+), New York ocean haul seine (ages 3–13+), Northeast Fisheries Science Center spring inshore survey (ages 3–13+), and three age-aggregated trawl indices from Connecticut (ages 2–6), New Jersey (ages 2+), and Delaware (ages 2–7). Juvenile surveys produce indices of young-of-year (age 0) in Maryland, Virginia, New York, and New Jersey as well as age 1 indices for Maryland and Long Island, New York.

The fishery-dependent indices represent a mix of commercial and recreational fishing data, including the Massachusetts commercial catch per trip (ages 7–13+) and the Connecticut volunteer angler catch per trip (ages 2–13+). Altogether, these data sources represent 55 age-specific indices that are used in the ADAPT–VPA model.

The MRFSS estimates of catch rate are not used in the striped bass assessment, although the time series of these catch rates have been compared with trends from VPA by the technical committee and found to be similar in recent years.

Economic and Social Information

Amendment 6 of the Interstate Fisheries Management Plan for Atlantic Striped Bass states that the Atlantic Coast Cooperative Statistics Program will require the collection of baseline social and economic data on all recreational fisheries through add-ons to existing recreational catch and effort surveys (Atlantic States Marine Fisheries Commission, 2003). This information is defined in the Atlantic Coast Cooperative Statistics Program documents as follows:

> Economic information includes information on market conditions in commercial fisheries (price and value information), as well as complementary information on recreational fisheries. Social sciences information is typically broader sources of information specific to commercial and recreational fishermen, their families, and the fishing community in general. For many managed fisheries economic and social sciences information is not available and is provided in an informal manner by fishermen during public comment periods. At times, this information is viewed as anecdotal and may be difficult to use in the fishery management decision-making process. (Atlantic States Marine Fisheries Com-mission, 2003)

Appendix D

Acronyms

ACCSP	Atlantic Coast Cooperative Statistics Program
ASA	American Statistical Association
ASAP	age-structured analysis program
ASMFC	Atlantic States Marine Fisheries Commission
BRD	bycatch reduction device
CCS	Catch Card Survey
CPFV	commercial passenger fishing vessel
CPUE	catch per unit effort
CRFS	California Recreational Fisheries Survey
DAMARS	Designs and Models for Aquatic Resource Surveys
EIA	Energy Information Administration
EPA	Environmental Protection Agency
FHS	For-Hire Survey
ft	feet
GIS	Geographic Information System
GMFMC	Gulf of Mexico Fishery Management Council
GPS	Global Positioning System
HPUE	harvest per unit effort

IMPLAN	Impact Analysis for Planning Model
in	inch
IRG	independent research group
kg	kilogram
lbs	pounds
LPS	Large Pelagic Survey
m	meter
MMS	Methodology, Measurement, and Statistics Program
MRFSS	Marine Recreational Fisheries Statistics Survey
MSE	mean square error
mt	metric ton
NOAA	National Oceanic and Atmospheric Administration
NMFS	National Marine Fisheries Service
NRC	National Research Council
NRCS	National Resources Conservation Service
NRI	National Resources Inventory
NSF	National Science Foundation
ORBS	Oregon Recreational Boat Survey
OSP	Ocean Sampling Program
OY	optimum yield
PacFIN	Pacific Coast Fisheries Information Network
PCS	Party Charter Survey
PFMC	Pacific Fishery Management Council
PSMFC	Pacific States Marine Fisheries Commission
PSSP	Puget Sound Sampling Program
RDD	random digit dialing
RecFIN	Recreational Fisheries Information Network
RTC	Recreational Technical Committee
SEAMAP	Spatial Ecological Analysis of Megavertebrate Populations
SEBS	Shore and Estuary Boat Survey
SEDAR	Southeast Data, Assessment, and Review
SEFSC	Southeast Fisheries Science Center

STAR	Science and Technology to Achieve Results
STAR Panel	Stock Assessment Review Panel
STARMAP	Space-Time Aquatic Resources Modeling and Analysis Program
SWHS	Sport Fish Statewide Harvest Survey
TED	turtle exclusion device
TIP	trip interview program
TPWD	Texas Parks and Wildlife Department
VPA	Virtual Population Analysis
VDTS	Vessel Directory Telephone Survey
VTR	Vessel Trip Report
WDFW	Washington Department of Fish and Wildlife